ODDITIES OF THE MINERAL WORLD

ODDITIES OF THE MINERAL WORLD

William B. Sanborn

with photographs by the author

VNR VAN NOSTRAND REINHOLD COMPANY
New York Cincinnati Toronto London Melbourne

This volume is dedicated to my four girls—my wife Joan and my daughters Debbie, Patti, and Leone—who have good-naturedly tolerated a determined collector for many years both at home and on countless field trips.

Library of Congress Catalog Card Number 76-6813
ISBN 0-442-27346-0

Printed in the United States of America
Designed by Loudan Enterprises

Published in 1976 by Van Nostrand Reinhold Company
A Division of Litton Educational Publishing, Inc.
450 West 33rd Street
New York, NY 10001

Van Nostrand Reinhold Limited
1410 Birchmount Road
Scarborough, Ontario M1P 2E7, Canada

Van Nostrand Reinhold Australia Pty. Ltd.
17 Queen Street
Mitcham, Victoria 3132, Australia

Van Nostrand Reinhold Company Ltd.
Molly Millars Lane
Wokingham, Berkshire, England

Library of Congress Cataloging in Publication Data

Sanborn, William B
Oddities of the mineral world.

Bibliography: p.
Includes index.
1. Mineralogy—Miscellanea. I. Title.
QE399.2.S26 549 76-6813
ISBN 0-442-27346-0

ACKNOWLEDGMENTS

Grateful appreciation is expressed to the following for permission to abstract directly or indirectly from various articles and books written by the author: *Natural History* (American Museum of Natural History); *Pacific Discovery* (California Academy of Sciences); *Sunset Books* (Lane Book Company); *Yellowstone Nature Notes* (National Park Service); the *New York Times;* and, especially, *Gems and Minerals.*

Special appreciation is offered to Dr. Gene and Manetta Tribbey, Dr. Peter and Virginia Bancroft, Dr. S. R. Silverman, Si and Ann Frazier, Wayne and Dona Leicht, Bill and Karla Larson, Ralph and Mimi Merrill, Steve Sinotte, Jack and Leslie Halpern, Frank and Helen Rentchler, Rosemary Luxton, Billie Casey, Loretta Cannon, Al Walters, and many other collectors, curators, and dealers who have encouraged my interest in the mineralogically odd and unusual.

C-1. Gem-quality tourmaline crystals from various mines in San Diego County, California are often found distinctively *zoned*. Two to four different colors may appear within the same crystal. The green-and-pink specimen on the right is from Pala, and the other four are from Mesa Grande.

C-2. Among the most prized of all unusual specimens are the handsome *twisted* smoky-quartz crystals from Switzerland. This fine example is flawlessly clear and, when viewed along the c-axis, shows a marked pitch or twist, somewhat like a propeller. The specimen measures 2½" × 1½" × ¾" and came from Göschenen, Canton Uri.

C-3. Agate nodules, or thundereggs, when cut and polished, may often show remarkable pictures, landscapes, or scenes. This astonishing seascape is complete with horizon, ocean, clouds, waves, sand bars, beach, and lagoon. It is half of a polished "egg" from the Chocolate Mountains, Imperial County, California. The scenic features are simply the product of inclusions of various mineralogical impurities deposited with the original chalcedony (agate) at various times as the nodule was being formed.

C-4. Many minerals can occur in *stalactitic* form. This lustrous, translucent aragonite stalactite is 10" long and is colored a strong aqua shade by inclusions of copper minerals. The specimen came from the roof of a small crystal-lined pocket, or vug, in a mine at Tintic, Juab County, Utah in the late thirties. It is a choice specimen due to the unusual coloring.

C-5. These are excellent examples of *inclusions* of one mineral within another. From top to bottom: papagoite in quartz, Messina, North Transvaal, South Africa; rutile in quartz, Tavetsch, Canton Grisons, Switzerland; stibnite in quartz, Manhattan, Nye County, Nevada. Inclusion specimens may be very attractive and constitute choice collector's items.

C-6. Six distinct *phantoms* can be seen rising like ghostly pyramids one on top of the other within this flawless quartz crystal. Phantoms are inclusions caused by impurities deposited on certain faces during different periods of the crystal's growth. This specimen is 2½" × 2½" × 3" and came from Belo Horizonte, Minas Gerais, Brazil.

C-7. For years Germany, Brazil, Uruguay, India, and Mexico have been the best sources of natural beautifully colored banded agates. At the same time artificially colored or dyed agates have been sold by the jewelry and tourist trades. This fine specimen is very unusual for an agate from North America, since it is *not* artificially colored. It is a polished slab cut from a melon-size agate nodule found in glacial debris high in the Beartooth Mountains of southern Montana.

C-8. The small Apex mine, near St. George, Washington County, Utah, has produced some of the world's most attractive *pseudomorphs*. This specimen is an azurite (blue) and malachite (green) pseudomorph after a group of selenite crystals. These specimens were not found in abundance. They are light in weight and very fragile.

C-9. Among the most common of all pseudomorphic-type mineral specimens are malachite pseudomorphs after azurite. In this case green malachite has completely replaced deep blue crystals of azurite, a similar copper-carbonate mineral. This specimen is 1½" × 1½" × 1" and came from the early workings of the Copper Queen mine, Bisbee, Cochise County, Arizona.

C-10. These are good examples of pseudomorphic shells, or *encrustations*. The specimen on the left is a 4" × 3" × 1" wafer-thin encrustation of brilliant quartz, calcite, and stilbite after a cluster of what were probably anhydrite crystals. It came from the traprock formations near Paterson, New Jersey. The specimen on the right is an odd geodal calcite-and-duftite encrustation pseudomorph after a 3" × 3" × 2½" cavity formerly lined with dolomite crystals. It is extremely thin and delicate and was found in the mines at Tsumeb, South West Africa.

C-11 and C-12. One of the most startling phenomena of the mineral world is the property of *luminescence*. Certain minerals from selected localities will turn brilliant colors under both short- and long-wave ultraviolet-light radiation. Turning color during exposure is called *fluorescence*, and glowing after the source is turned off is called *phosphorescence*. This is a typical specimen from localities around Franklin and Sterling Hill, New Jersey. At the left the specimen appears in ordinary daylight, and at the right under short-wave ultraviolet rays. The green is willemite, the cerise calcite, and the black franklinite. This specimen also phosphoresces brilliantly.

CONTENTS

Chapter 1

INTRODUCTION

This book is not a basic introduction to the world of minerals, fossils, and lapidary arts; a book on the chemistry of mineralogy; a pedantic treatise on crystal morphology; an encyclopedia of localities; a photographic presentation of outstanding mineral specimens and gemstones; a compendium of field trips; nor a field guide to mineral identification. There are a number of excellent, well-written, and profusely illustrated books available on these and related topics. Producing another such book would be a classic example of reinventing the wheel. Consequently, you may wonder what this book is all about.

It is essentially designed to assist the interested layman, the beginning or advanced amateur mineral or fossil collector, and the lapidary enthusiast in understanding *some* (not all) of the more unusual phenomena that occur in the fascinating world of mineralogy. Some of the fundamentals of descriptive mineralogy are included. In some cases there is a degree of oversimplification for the reader who is not familiar with complex geologic, mineralogical, or crystallographic terms.

This book is an interpretive and, hopefully, stimulating *sampling* of the spectrum of provocative phenomena within the field of mineralogy. A wide-angle approach is used, since not only are minerals per se discussed but also some tangential aspects of mineralogy involving speleology (the study of caves), paleontology, and the lapidary arts. For example, both mineralogists and paleontologists are apt to forget that a fine specimen of colorful petrified wood, a pyritized brachiopod, or a silicified insect involves a combination of both sciences. The lapidary artist who fashions a beautiful pendant from agatized coral is working with paleontological material.

The processes, phenomena, and specimens discussed in this book range from such fascinating items as stalactites, thundereggs, geodes, and fossil wood to more sophisticated specimens such as insect-centered concretions, pseudomorphs, and tektites. What is odd and unique to one person may not be to another; consequently, the wide variety of coverage. The unusual ranges from the novice's amazement at a fossil pine cone and the confusion of stalagmites and stalactites to Switzerland's twisted crystals of smoky quartz. Hopefully, there is something for everyone in the following chapters, and once these vignettes are put together, attention is focused on mineralogy, one of the most fascinating and varied of all sciences.

This book does not deal in depth with the exceptionally rare one-of-a-kind museum or private-collection oddities. Volumes could be devoted to the hun-

C-1

C-2

C-3

C-4

C-5

C-6

C-7 C-8

C-9

C-10

C-11

C-12

dreds of unique specimens of odd mineral combinations, unusual format, unique size, or simply odd occurrences at a certain locality. There will always be many of these, which, of course, adds a constant atmosphere of anticipation and excitement to collecting.

One of the most interesting aspects of the field is the amazing variety of collection specialties contained within minerals, fossils, and the lapidary arts. A reasonable insight into this factor is discussed, with emphasis on how valuable fine, well-organized collections of materials may be. The subject of values is one of the more unusual considerations in this field. Consequently, a discussion of the values, not only of mineralogical oddities but of minerals in general, is presented in the last chapter. Values are a composite of several factors.

You may see specimens of the type covered in this book in museums or in private collections, and they prove completely fascinating, yet very puzzling, unless accompanied by adequate explanation. Many of the items covered in the subsequent chapters are mentioned only briefly in other publications, but here the collector will find useful data explaining some of the most unusual and interesting facets of mineralogy.

There are various aspects of mineralogy upon which the opinions of noted experts differ. Similarly, many terms have different spellings, particularly localities. Specific localities may also change as the years go by. Old mines are sold or new mines are developed in old mining areas, and the names may change. Similarly, locality misinformation or outright falsification is a well-known factor to experienced mineral collectors. In addition, both the accessibility and the availability of materials at localities are flexible and subject to change. Some localities simply become exhausted. Thoughtless vandalism has caused many fine localities to be closed to everyone.

I hope that these chapters open up new vistas of understanding, appreciation, and research for the hobbyist. They certainly indicate that there are many aspects of the earth sciences which remain clouded or even unexplained. Much more research and subsequent interpretation is needed in many areas. Like a fine painting, a crystal, fossil specimen, or beautiful agate should be enjoyed, and the more you know about a specimen's story, the greater your understanding and subsequent pleasure.

Enjoy!

The Specimens

The specimens that appear in the illustrations are typical examples of the phenomena discussed. With a few exceptions, they are not rare, one-of-a-kind samples but rather examples of the type of specimen that might readily be found or purchased by an average collector. They are representative of some of the more unusual yet reasonably common occurrences of a given phenomenon.

There are, of course, hundreds of exceptionally spectacular individual mineral specimens from selected localities that have become well known and widely pictured. Many of these are, indeed, odd but represent such unique finds that the average collector would have to be content with viewing such specimens in a museum; consequently, specimens of this type are not included. Similarly, unusual "combinations" of crystallized minerals, such as garnet in galena, apatite with wolframite, etc., and various unusual crystal forms, such as quartz Japanese twins, are not covered, since understanding and appreciating their chemistry and crystallography would involve a reasonably sophisticated undertaking. As indicated, there are fine books already in print on such subjects.

The Photographs

The photographs are designed to show the minerals in an unstaged, see-it-like-it-looks manner. No attempt has been made to achieve tricky or dramatic lighting or to create false internal illumination. Such procedures can produce very beautiful photographs, yet they may also greatly distort or exaggerate the actual appearance of the specimen.

Keep in mind that this volume deals only with *some* of the odd or unusual specimens of the mineral world. A few of the illustrations are of very attractive or showy specimens, such as the malachite pseudomorphs after azurite. However, many specimens discussed are only interesting in terms of the story they tell—they are assuredly not attractive! The cone-in-cone specimens are a good example—aesthetically uninspiring, yet their explanation is provocative.

The film used for the color illustrations is 35mm Kodachrome 25, daylight type, ASA 25; the black-and-white shots were done with Plus-X Pan, ASA 125, size 120 and Ilford Pan-F exposed at ASA 80, size 135.

Both of the cameras used were of the single-lens-reflex type: a 120 Hasselblad 500-C with the 80mm Planar f/2.8 and 120mm S-Planar f/5.6 lenses, and a 35mm Nikon FTN equipped with a 55mm f/3.5 automatic Micro-Nikkor lens. Lighting was direct and simple in order to highlight the basic features discussed. Photographs not shot under daylight conditions used two BCA-No. 81 *blue* photoflood-lamps in 7½" reflectors. A time-exposure technique using a small lens aperture and fill-in with flashbulbs was used for the cavern pictures. The backgrounds are swatches of various-colored felt, matte-finish construction paper, or sand.

Incidentally, although I use many lenses for different purposes, the 55mm Micro-Nikkor, or a similar lens, is an excellent all-around choice for natural-science photography, both in the laboratory and in the field. This exceptionally flexible standard lens enables the photographer to shoot any normal scenes and to do extremely small closeups *without* using a bellows, extension tubes, or add-on lens components. Also, by appropriate adjustment of the shutter-speed dial or the f/stop opening ring, the exposure is automatically computed through the lens and finder on this type of through-the-lens metered camera. Although the lens only has a maximum aperture of f/3.5 and would be considered "slow" for some types of work, it is completely adequate for normal or closeup exposures, since you can easily compensate with a fast-ASA film and appropriate shutter adjustments. With the excellent fine-grain color and black-and-white film emulsions available today, an f/3.5 lens should not be considered a problem at all. For studio shots, the camera was mounted on a very heavy, professional-type tripod, and all exposures were tripped by cable release. Most photoflood exposures were at ½ or ¼ second.

Chapter 2

COLLECTING CRYSTALS, MINERALS, AND FOSSILS

"Hey, Dad, what's this?" I asked that question some forty years ago when I picked up what I later learned was a small garnet crystal from some crushed gravel in our driveway. I rolled the tiny dodecahedron around in the palm of my hand. It flashed ruby-red in the sunlight, and the precision of its beautiful shining faces was fascinating. That little garnet did it—I've been an avid collector of crystallized minerals ever since.

Although people who look for and collect all types of rocks, minerals, fossils, and lapidary materials are broadly referred to as "rockhounds," there are many more sophisticated levels of these popular pastimes than this term implies. Those who collect minerals may do so on a very casual, fun basis or pursue the hobby as a serious, sophisticated, advanced collector. The fossil enthusiast may well evolve into an expert amateur paleontologist. The lapidary devotee may readily become a first-class lapidist, gemologist, or silver-goldsmith.

In recent years all of these earth-science-oriented hobbies have shown positively astonishing growth on a worldwide basis. There are several reasons for this. These hobbies may be enjoyed by people of all ages who live in *any* part of the country—America is blessed with an abundance and an unsurpassed collection of major geological phenomena to visit and enjoy. They are superb family activities and can combine outdoors and indoors—outdoor for hunting and digging when weather permits, and indoor for sorting, preparing, cataloging, displaying, or cutting and polishing. The hobbies can be as high-priced or as inexpensive as one's pocketbook dictates and yet remain enjoyable. Once you become really interested, it is amazing how many specimens may be acquired locally just by learning where, when, and how to hunt and dig. Hobbyists can do most of their collecting within a radius of their homes, on weekend trips, or perhaps on vacations devoted to interstate travel in search of materials. Materials may also be obtained by trading with other collectors, at local mineral or lapidary shops, or at local, state, regional, and national shows.

Mineral collecting can be pursued to any degree you desire. You may enjoy simply collecting colorful specimens, or you may want to delve deeply into the chemistry and crystallography of minerals. You can create decorative displays for your home. There's even an economic aspect. Many experienced collectors regard outstanding specimens as good investments. Recently, for example, a superb private collection purportedly sold in the West for $400,000.

The earth's crust consists of rock, which in turn consists of various combinations of minerals. Almost all of these minerals, which are composed of a distinct chemical compound or a single element, occur in crystallized form under certain conditions. The rocks in which these crystals form are often cracked, fissured, or have natural openings in them, and it is in these pockets, seams, or cavities that the minerals may crystallize.

Minerals can be found almost anywhere that rocks are exposed—shallow river beds, ridges, gravel pits, quarries, road-construction sites, and building foundations. Old mine dumps where waste rock has been piled up are a rich source of minerals (but stay out of *abandoned* mines), as are seashore cliffs and lake shores in many localities.

Crystals may be embedded in the rocks in which they originally were formed, or they may weather out due to natural erosion and be found on the surface nearby. While loose crystals are fairly abundant in some localities, more often than not the crystals are found *in situ* in the rock or related cavities and must be carefully extracted with hammer and chisel. Obviously, mines are a great source of specimen material.

One of the most intriguing aspects of collecting crystallized minerals is that, unlike coins, stamps, or model trains, no two mineral specimens are ever exactly alike. Thousands of quartz crystals, for example, have been found in various parts of Arkansas, but no two are really identical. A specimen's size and quality, as well as its particular aesthetic format, are factors to be considered in establishing a fine collection.

The tools needed for mineral and fossil collecting are minimal and inexpensive. The only unusual tool is a mineral or rock pick, preferably one with a picklike point and one with a chisel point. These picks come in different sizes and weights to suit the variable swinging and pounding capacities of children and adults. Other useful tools, dictated by the type of locality involved, might include various sizes of cold chisels, a pry bar, a heavy cracking hammer, sledgehammers, shovels, and even sifting screens. Appropriate carrying bags and packing material, such as newspaper, tissue, cotton, or even egg boxes, are essential. The tool investment is small—the enjoyment is enormous!

Getting Started

How do you get started in the mineral, fossil, or lapidary field? It is really quite easy. One of the best ways is to join one of the many local clubs, guilds, or societies found in almost every major community, county, or region. There are over 1,500 well-organized clubs in the United States alone, with a membership close to 100,000. Clubs or societies will differ in depth of organization, varying from very informal, loosely knit groups to others that are exceptionally well-organized, dynamic, and articulate. Most clubs have regular meetings and frequent field trips and often sponsor or cosponsor annual shows as well as conducting various classes, as in the case of lapidary-oriented groups. There are general-interest clubs and clubs that are very specialized, in such fields as fluorescent minerals or micromounts. Information about local clubs and organizations is usually readily available from local rock or lapidary shops; look for them in the Yellow Pages under "Lapidary" or "Minerals."

If you cannot locate a society locally, write to the offices of the American Federation of Mineralogical Societies, Att. AFMS Secretary, at 4727 N. 24th Street, Arlington, Va. 22207, and they will gladly supply the name, address, and nearest contact for a society in your area.

As indicated earlier, there are dozens of excellent publications on just about any aspect of these hobbies you can imagine. Libraries are an obvious source for booklets, books, or periodicals. The next best source is your local mineral or lapidary shop, and the proprietors are always anxious to assist hobbyists. There are hundreds of such firms throughout the country. For both the novice and the experienced mineralogist there are dozens of good guide books on localities and how-to-do-it booklets handled by mineral shops. Some are organized on a regional basis for field collecting and give specific details on collecting in that area, plus locality maps and helpful collecting pointers.

Subscribing to publications such as *Gems and Minerals*, *The Lapidary Journal*, or *The Mineralogical Record* (see bibliography) is very worthwhile, not only for the articles but also for the advertisements. A vast amount of interesting and useful data has been published in these three particular journals over the years.

In addition to participation and reading, there is really no better way to learn about minerals, fossils, and lapidary materials than to *see* them: viewing other collectors' collections and visiting museums is recommended highly. There are many remarkable mineral displays in the United States and Canada. Some of the best are:

The United States National Museum (Smithsonian), Washington, D.C.
The American Museum of Natural History, New York, N.Y.
The California Academy of Sciences, San Francisco, California
Cranbrook Institute of Science, Bloomfield Hills, Michigan
Colorado Bureau of Mines, Denver, Colorado
Harvard Mineralogical Museum, Cambridge, Massachusetts
Chicago Natural History Museum (Hall 35), Chicago, Illinois
South Dakota School of Mines and Technology, Rapid City, South Dakota
University of New Mexico Geology Museum, Albuquerque, New Mexico
Montana School of Mines, Butte, Montana
The Royal Ontario Museum of Mineralogy and Geology, Toronto, Canada

The displays at the United States National Museum (Smithsonian), The American Museum of Natural History, and the Harvard Mineralogical Museum are particularly magnificent in terms of the quality and variety of superlative materials on display.

Several of the world's finest private mineral collections are in the United States. Access to these is best established via mineral clubs and advanced collectors, who in turn may arrange visits to other private collectors for the sincerely interested. An organization of advanced mineral collectors, many of whom own outstanding private collections, is The Friends of Mineralogy, a group associated with *The Mineralogical Record.* Incidentally, an important rule of etiquette when examining another hobbyist's collection is not to handle the specimens without permission.

Collection Categories

The beginner may be somewhat confused by the various *types* of collections maintained by hobbyists in the earth-science field and also by the variety in *size* of specimens featured in a collection. Since odd and unusual mineral specimens may be found in all kinds of collections, one of the most interesting aspects of the field is the great *variety* of well-defined collection formats and specialties involved.

When you begin collecting, your approach will usually be casual—in other words, you will probably start a collection composed of almost anything and of any size. There is nothing wrong with this, but soon a metamorphosis will begin to take place. The more you learn, the faster you will realize that there is really a lot of practical common sense involved in focusing on a particular size and type of collection. There are several factors involved in specializing in a particular size of specimen, and almost any collection assumes a more desirable professional appearance if one basic size is featured.

One of the most enjoyable aspects of collecting minerals, fossils, or lapidary materials is sharing them with others in an attractive display. It is amazing how a collection of average specimens can appear to be of higher quality simply by virtue of good display techniques. There is nothing more distressing than to view someone's collection and find that it appears to have been arranged with a coal shovel at a distance of ten paces from the cabinets. Whether your display is on shelves, in drawers, or in glass cases, it should be enjoyable, not a disaster area.

Displaying and housing a collection are very important considerations that involve several variables. A small apartment can easily accommodate several thousand micromount-size specimens in neat cabinet drawers. How you choose to organize, house, or display your collection is up to you, but well-organized collectors focus on a method of display that is within their financial means and also fits well into the living room, den, workshop, or office space involved.

There are five size categories that are generally recognized in the hobby. Each has advantages and disadvantages.

Micromounts: These are tiny, crystallized specimens that are viewed under a binocular microscope. There are special micromount societies and rigid rules for appropriate mounting in tiny boxes for competition (if you desire). The fine material available is legion and low in cost, and perfection is usually attainable. Tiny micromount specimens can be of astonishing beauty, and excellent specimens are available in this tiny size that few could ever afford or obtain in any other size. There is also the marked advantage of being able to store a large collection in a very small space.

Thumbnails: This is a descriptive but clumsy term—nevertheless, one that has been in the jargon of the hobby for many years. It refers to a small specimen about the size of a thumbnail to 1″ × 1″. Specimens of this size do not necessarily require a microscope for viewing, and all the advantages of micromounts still apply: superb quality, relative abundance, low cost, compact display, and storage potential.

Miniatures: This is an extremely popular and practical size, generally referring to a specimen about 1″ × 2″, or egg size. These are large enough to make a very effective and dramatic display in a case or drawer, yet are still small enough to present favorable availability, purchase price, and storage considerations. Miniatures are usually housed in drawer cabinets, but a word of caution is needed: be sure you install a stop device on each drawer so that you do not inadvertently pull the drawer all the way out and dump the specimens on the floor. It happens frequently!

Cabinet: This is a long-standing, almost self-explanatory term referring to specimens particularly suitable for attractive cabinet display. Two additional terms are used within this category: *small cabinet*, referring to hand-size specimens; and *large cabinet*, involving specimens around 8″ × 10″. Well-selected cabinet specimens can provide a spectacular display, yet many choice specimens are very difficult to obtain or afford in this size. Superlative miniature specimens of wulfenite, for example, are readily available, but large cabinet specimens in fine condition are very costly.

Figure 2-1. There are various *sizes* of mineral specimens in which collectors may specialize. From top to bottom, left to right these fine crystallized specimens exhibit the following generally recognized sizes: large cabinet, cabinet, small cabinet, miniature, thumbnail (1" × 1"), and micromount. There are advantages and limitations to each size.

Museum: This term can refer to any top-quality specimen worthy of museum display regardless of size—a ½" diamond crystal or a 2-ton meteorite. The term, however, also refers to exceptionally large specimens that can practically be displayed only within a museum environment. For example, a 2′ × 4′ plate of huge quartz crystals from Arkansas, calcites from the Tri-State Mining District, or fluorites from Cave-In-Rock, Illinois probably are best displayed in a museum rather than a home living room!

Along with specimen-size decisions, specialization in a particular *type* of specimen enables an amateur to assemble an outstanding collection tailored to fit his individual interests and pocketbook. Many such privately assembled collections have great scientific as well as cash value. Specialization can also provide a stage for viewing the incredible variety of forms and oddities that a particular mineral or mineral family may manifest. A fine collection devoted only to calcite (an abundant mineral throughout the world) will reveal a spectrum of color, shape, format, and morphology that can be amazing.

Following are some typical type specializations favored by many collectors Even within these few categories there are many subspecializations, representing one of the most unique in-depth aspects of any pastime.

Rocks: Although many are prone to make jokes about collecting rocks, don't underestimate the complexity of this study. Rocks constitute both the basement and the attic of our planet. The scientific study of rocks is called *petrology.* There are countless hundreds of different kinds of rocks, the identification and classification of which are often determined by examining a polished microthin section under a microscope.

Ores: Ores are the fundamental ingredients of all mines, and consequently, the commercial value of ore-analysis technology is obvious. Ores, although usually very unattractive, can hold fascinating stories. There are collectors who specialize, for example, in gold ores, silver ores, and even ores of mercury.

Fossils: Within this category are many fine, specialized organizations devoted to the study of both vertebrate and invertebrate paleontology. Without a doubt, collecting fossils, coupled with an understanding of what they represent, is one of the world's most fascinating hobbies.

General collection: This is a broad term covering a collection involving a little of everything, in all sizes. Tastefully displayed and well cataloged, it is probably the most interesting of all collection types for the average person.

General collection with suites: A basic general collection will frequently show a particular specialized interest of the collector by virtue of suites of a particular mineral, specimens from one mine, etc.

Lapidary specimens, materials, and gemstones: There are thousands of enthusiastic hobbyists with fine collections of polished thundereggs, slabs of agate or petrified wood, polished gemstones in either cabochon or faceted format, and various attendant jewelry-making arts and crafts. The amateur is capable of thoroughly professional accomplishments.

Meteorites: Indeed a difficult specialty, but there are hobbyists who are entranced by the various types of meteorites and tektites. With very few exceptions, specimens must be purchased or obtained by exchange. The chances of finding a meteorite are incalculably slim.

Pseudomorphs: One of the most intriguing specializations, or suites within a collection, is pseudomorphs, minerals that outwardly resemble different minerals. The mineralogical story revealed by many pseudomorph specimens is fascinating, and learning to recognize them is even more so!

Inclusions: This is an unusual specialization featuring various minerals included within other minerals. For the student of mineralogy such specimens can be exceptionally stimulating and often pose some disturbing problems!

One mineral: This is a very popular specialization focusing on one particular mineral from various worldwide localities. Collections of some of the more common minerals, such as quartz, calcite, barite, or fluorite, can be breathtakingly beautiful and fascinating in terms of variety.

One mine or mining district: Usually this is represented by extensive suites within a basic mineral collection and constitutes a particularly valuable scientific approach. Collectors may specialize in the minerals of the Mammoth-St. Anthony mine, the '79 mine, minerals of New Mexico's Magdalena Mining District, or specimens from the fluorite mining district around Rosiclare and Cave-In-Rock, Illinois.

One county, state, region, or country: A self-explanatory category, but one frequently encountered, such as a collection of fine minerals from San Bernardino or San Diego County, California; a collection focused on specimens from Michigan or Nevada; or the minerals of New York City and its environs. Country collections are very popular, especially in shows, and might feature specimens from South West Africa or the continent in general, Brazil, Mexico, Canada, or Australia—all of which produce a particularly fine assortment of superb specimens.

Single crystals: This is a fine scientific specialization for those who become interested in crystallography. A fine collection of choice natural single crystals can be almost unbelievable in terms of the varied geometry and perfection involved. It can be a very sophisticated study.

Luminescent minerals: Special societies and clubs devote their efforts to collecting, displaying, and studying the phenomena of luminescent mineral specimens that fluoresce or phosphoresce.

Dana collection: This is a scientific mineralogical specialization aimed at collecting as many as possible of the various species of minerals listed in the *Dana System of Mineralogy.*

One mineral family: This very rewarding specialization provides somewhat more depth than collecting a single species. Here, for example, the collector might focus upon iron, lead, or copper minerals; there are dozens of different minerals whose basic formula involves Fe, Pb, or Cu.

General chemical classification: This classification is generally pursued by an advanced collector either as a collection emphasis or as an extensive suite within a basic collection. For example, emphasis might be upon minerals that are carbonates, phosphates, silicates, borates, sulfates, oxides, or arsenates. Those with a flair for chemistry find this rewarding.

Gemstone minerals: This is a fascinating collection, but often difficult to assemble. Nevertheless, there are many outstanding museum displays and private collections devoted to specimens of minerals used for jewelry or ornamental purposes. If they are combined with examples of the finished product, such collections can be spectacular.

Attractive minerals: This may sound somewhat obvious, but it is a very common type of specialization! Many collectors are fascinated by the attractive minerals, and a well-selected display case can be the highlight of a home. Fine, attractive specimens have high aesthetic appeal.

Specimens personally collected: This often exists as a suite within a basic collection or surfaces in displays at mineral shows. It is always encouraging and stimulating to see precisely what someone has actually found!

A mineral collector soon learns that there are a great number of different minerals in the world, and he can become familiar with perhaps fifty to one hundred with modest but studious efforts. To those who become hooked on mineral collecting, the availability of specimens becomes more intriguing. New species are, in fact, added to the list of known minerals each year. This is the result of new mines in heretofore nonmined areas, new ore bodies mined in old and well-known mining districts, and detailed scientific analysis of existing specimens. There are approximately two thousand different mineral specimens that have been identified at present. Of this number perhaps two hundred might be considered common, another three hundred fifty uncommon, and the remainder scarce to rare. There are many mineral species that are extremely rare, and the number of known specimens may be limited to a dozen or so samples.

Not all minerals are attractive or occur in a well-crystallized form by the remotest stretch of the imagination! Many common or rare species may consist of only a powdery substance on a rock matrix, a dull encrustation that is hardly discernible, or a crude, imperfect crystal. There is absolutely no correlation between rarity and beauty (or oddity) in the mineral field. The most common crystallized mineral in the world is quartz, and it may be found in drab, completely unattractive crystals or in specimens of spectacular beauty. At the same time a very rare and complicated borate mineral, rhodizite, is prized by advanced collectors but occurs in small, unattractive, colorless to yellowish crystals. However, some rare minerals may occur in attractive and unusual crystals, adding another dimension to the specimen. A good example would be the small but fascinating indigo-blue crystals of the lead-copper mineral cumengite found at Boleo, Baja California Sur, Mexico. This very rare mineral occurs as remarkable parallel overgrowths on crystals of boleite and pseudoboleite, encrusting the other crystals in such a manner as to simulate twin crystals. This is a typical example of how sophisticated and intriguing the study of mineralogy can be. The little crystals of cumengite, boleite, or pseudoboleite are much prized by advanced mineral collectors.

The fact that a given mineral—for example, wulfenite or smithsonite—may be found in an astounding variety of colors and crystal habits from various localities throughout the world also adds a truly unique dimension to the hobby of mineral collecting. It is a hobby that flourishes on fine, old classic specimens or "old-timers" as well as on the latest in mineralogical finds. It is never static!

"Solid as a rock" is a complete fallacy if you think that rocks and minerals are permanent. No rock lasts forever, and most rocks are assuredly not solid, since given sufficient time all will change or disappear via oxidation or erosion, and solutions do flow quite readily within many rocks. Consequently, rocks and minerals are really not as stable as you might think, and many are *very* unstable. Once they are exposed to air or various solutions within the earth's crust, minerals may go through marked changes. Ask any knowledgeable mineral collector and you'll find that some specimens of marcasite are very unstable, and a specimen that was very showy five years ago may suddenly disintegrate into a pile of gray, sulphurous-smelling dust. Even light can radically affect certain silver- and arsenic-base specimens that are photosensitive. Gorgeous translucent ruby-red crystals of proustite must be kept in the dark or they will accumulate a dull gray opaque surface due to the effect of light on their silver content. Similarly, fine crystals of orange orpiment will turn to yellow powder if exposed too long to both air and light. Many collectors do not include certain borate minerals within their collections due to the difficulties in keeping them from oxidizing and turning to dust.

One last thought in this introduction to collecting is the importance of *localities*, especially as they pertain to fine mineral specimens. One key to a well-managed and potentially valuable collection is adequate and accurate labeling. Various localities have produced crystallized specimens that are highly distinctive, and since they may eventually be exhausted, labels are important. Knowing precisely where a specimen came from definitely adds to the specimen's value.

Most professional labels include the name or names of the minerals involved, name of the mine if known, mining district, county, state, and country. Purchased specimens should always be accompanied by a label furnishing as much data as is available. Never throw away a label—keep it with your own. Most collectors treasure the labels that accompany a fine old-timer, thereby tracing the history of ownership of a specimen.

Obtaining accurate locality data is often difficult, since new finds are frequently purposely mislabeled to keep the competition from finding the precise site! Localities in Brazil, Mexico, and Africa have been subjected to this subterfuge for years. It might be said, somewhat facetiously, that trying to keep up with new country names plus the accompanying renaming of cities around the world is almost a hobby in itself!

Being able to recognize localities can be just as important and just as much fun as recognizing the minerals themselves. Localities are the other half of the art of building a fine mineral collection.

Chapter 3

UNUSUAL NATURAL CRYSTALS AND OTHER MATERIALS

What is a crystal? The experienced collector, curator, or dealer may forget what a tremendous curiosity a fine group of natural crystals is to someone who knows little about rocks and minerals. Crystals are assuredly among nature's greatest achievements in natural beauty, symmetry, and color. To the neophyte, it seems almost unbelievable that a beautiful, brilliant, sharp-faced group of quartz crystals just "grew that way" and that man has not done any cutting and polishing! Perhaps this is one of the reasons why collecting crystallized minerals has become so popular—the sheer fascination of the beauty, chemistry, and geometry involved.

An interesting question at this point is: "What is the most common mineral on the earth's surface?" Most people would quickly answer, "quartz," but this is wrong: quartz is second. Water is a mineral, the most abundant species on the surface of our planet. Water is one of the few minerals that is liquid at room temperature. (Mercury is another.) This unique mineral exists as a solid, ice; a liquid, water; and a gas, steam. It crystallizes in the hexagonal system. Water *is* listed in all professional, scientific mineralogical studies and may be found in *Dana's System of Mineralogy*, vol. 1, 7th edition, page 494.

Basic Characteristics of Crystals

Before reviewing a few of the more interesting aspects of crystals, some fundamentals are in order. A crystal is a very special, natural, multisided form highly characteristic of each individual mineral. Now, what is a mineral? One definition (although a bit complex) is: any chemical *element* or *compound* occurring as a natural substance that is part of the earth's crust. It has a characteristic but varying chemical composition and an essentially orderly atomic structure. Under ideal formation conditions this internal atomic structure is represented by the external appearance of the mineral as a crystal. Crystals are marvelously complex and might be considered the visible manifestation of the atomic structure.

Rocks, in turn, are composed of aggregates, or combinations, of one or more minerals; igneous, plutonic, sedimentary, and metamorphic are the basic categories. Generally speaking, *igneous* and *plutonic* rocks were at one time molten material, which was extruded either on the earth's surface or as deep-seated masses that cooled slowly far beneath the earth's crust. *Sedimentary* rocks are composed of the coarse to fine, eroded and recemented fragments or particles of other rocks, which may have accumulated as sediments in ancient deltas, lakes, sea bottoms, or even sand dunes. *Metamorphic* rocks are either sedimentary, igneous, or plutonic materials that have been altered in some manner by pressure and heat; the change does not involve melting. There are hundreds of different complex classifications of rocks according to the type and percentage of minerals involved in a given specimen. This is the subject matter of petrology and petrography. However, rocks are the host. They provide the environment in which crystals may form.

The term *ore* refers to an extensive deposit of a given rock or mineral having commercial value. For example, rocks in which the elements gold, silver, zinc, and copper occur in sufficient quantities to warrant mining are called ores of those respective minerals. However, the term ore does not necessarily mean a metal-bearing rock. Ores from which borax, gypsum, and potash are extracted may be a very lightweight rock or even powdery or cakelike residue materials obtained from the evaporation of hot water. As a matter of fact, a limestone quarry might be considered a source of an ore of cement. An extensive ore deposit of commercial value is often called an *ore body*, and each such concentration throughout the world is quite different from the others. They occur in different types of rock, containing different percentages of various minerals, and consequently provide the source of many different crystallized specimens. Crystallized mineral localities always offer a characteristic individual "look" about the specimens that differs from other localities.

Mineral species, of which there are some two thousand, differ from one another by virtue of the different chemical elements and the proportions thereof that they contain. The atoms of these combinations, or compounds, normally assume a characteristic orderly pattern. This specific organization in turn identifies an individual mineral species. Technically, a crystal is the idealized physical representation of a mineral.

Of the millions of tons of rock in the ore bodies of the world, only a few tons are in the form of fine crystal specimens. The most common physical state by far of all minerals is a solid, granular, or rocklike form of the element or compound. This is referred to as the *crystalline* form of the mineral and is essentially composed of millions of tiny crystals in a jumbled aggregate. Under ideal conditions a single *crystal* of a mineral may form. A crystal begins life within specific parameters controlled by the physics and chemistry of the mineral involved. It starts with an initial alignment of its atoms and repeats itself in an orderly pattern so that the atoms are in virtually identical alignment from one end of the crystal to the other—much like stacking blocks or bricks.

The finest crystals will occur where the best type of environment has been provided. This is a combination of an opening (cavity, vug, crack, pocket, or fissure in the host rock), hot or cold liquid solutions containing mineral material, and various degrees of heat and pressure. Under these conditions the external development of crystal faces may reach perfection. These are all variable factors and may differ considerably between localities. There are other types of crystal environments as well, such as dry lake beds, muds, and silts; around hot springs and fumaroles; the lining of geodes; or even within formerly molten rocks that cooled very slowly, allowing the mineral to develop as crystals in the solid mass. Later these may erode at the earth's surface, with the softer materials disintegrating and leaving harder materials behind as well-developed crystals. This is frequently the case with feldspar.

Almost all minerals have been found crystallized, but there are some exceptions. An often used description, although it is somewhat trite, is that a fine crystal might be referred to as a "flower of the mineral kingdom." This is quite accurate, since a crystal of a mineral is perhaps its most distinguishing identification factor—like a fingerprint. Given a good crystal, small or large, a mineralogist can usually identify the specimen. Other identification techniques include chemical or x-ray laboratory analysis, which in both cases yields a precise statement of the atomic structure and its proportions and, in turn, the name of the mineral species.

Crystals may grow to any size, depending on the original environment. Perhaps the most perfect of all crystals are the tiny ones best observed through a binocular microscope. Other crystals can be as large as an automobile—and mines have been opened in monstrous single crystals of beryl and calcite.

Crystals are found as single units, clusters, or extensive coatings lining openings in a host-rock matrix. Crystal specimens may be found that show only one mineral on the specimen, such as a plate of quartz crystals. However, multiple crystallizations are particularly common in ore bodies and can produce two or more minerals crystallized on the same specimen. For example, many of the mines in the old United States locality referred to as The Tri-State District have furnished beautifully crystallized specimens showing calcite, galena, sphalerite, and chalcopyrite on the same specimen. As is often the case, multiple crystallizations represent different stages of mineral deposition in a cavity: some minerals were deposited before others as initial solutions disappeared and were replaced by different ones. It is also possible for two or more minerals to crystallize simultaneously in a pocket.

It should be noted that a mineral found in a crystalline form in an ore body will not necessarily be found in fine crystals. The right conditions for crystallization may simply have not existed when this particular rock mass was forming. All ore bodies differ. For example, many gold mines in the foothills of California's Sierra Nevada were located in huge masses of pure quartz, yet in many of these no quartz *crystals* were ever found.

Quartz is also a good choice to illustrate the wide difference in appearance associated with a given mineral. Crystallography sounds very precise—and it is—but within this precision is flexibility. Quartz crystals from several different worldwide localities may at first appear to be superficially different. For example, they may be opaque, water-clear, filled with various inclusions, or lavender, black, pink, or reddish in color. The crystals may be microscopic or enormous; terminated prisms or doubly terminated; long and thin or short and fat. Nevertheless, regardless of the particular color, shape, or habit in which the quartz is found, there will always be a set of common denominators present: the number of faces on the crystal and the size of the angles between the faces. This particular set of denominators is the hallmark of quartz and would be true no matter where the specimen came from.

Not infrequently crystals may be *modified* in an odd or unusual manner. This may mean that the normal crystal-face format is markedly disproportionate, elaborated, or distorted. Quartz again is a good example, since some quartz crystals may need considerable orientation to discern the well-established pattern of facial alignment. Quartz can mimic topaz, beryl, and even calcite. Another modification involves a specimen that may show two different crystallization habits occurring on the same crystal, perhaps complicating easy identification. This occurs frequently with galena, fluorite, and calcite. Most mineral species that are found well crystallized are also found in a fascinating variety of modifications.

Following closely on the heels of modification in terms of the unusual is *twinning.* Most crystallizing minerals may occur as pairs, or twins, whose growth has been the product of very complicated and highly distinctive twinning laws applicable to the mineral. Learning to recognize true twinned crystals as opposed to two merged singles can be an interesting exercise!

Another interesting aspect of crystals is color. Minute impurities or variables within the basic chemistry of a mineral can produce a bewildering variety of colors. For example, the mineral fluorite is fairly common and is often found well crystallized in very attractive and choice specimens. Due to slight chemical variations fluorite occurs as fine specimens in dark green, blue, pink, transparent, lavender, purple, gold, yellow, apple-green, and aquamarine and frequently shows three zones of color in a single crystal; shape and angles between the crystal faces remain the same no matter where the specimen was found.

Mineral color is also subject to very misleading stereotypes. For example, if you ask someone, "What color is a garnet?", the answer will almost always be "red." Red garnets are the type commonly used in jewelry, since they are the most abundant in gemstone quality. However, fine, lustrous garnet crystals may be black, brown, white, olive, pink, green, salmon, yellow, orange, or a sensational cinnamon color. Obviously, specimens within the garnet series are quite versatile colorwise.

All mineral species crystallize within one of six fundamental crystal systems, determined according to the geometrical orientation of crystal axes: isometric, tetragonal, hexagonal, orthorhombic, monoclinic, and triclinic. Crystallography involves considerable mathematics, geometry in particular, and there are excellent introductory books on the subject. Much can be learned by studying at home or with fellow hobbyists or society members. Mineral collecting can be pursued to highly scientific depths, and this aspect proves stimulating to many collectors.

Twisted Crystals

There are several minerals which may show curved faces as an aspect of normal crystallization. For example, dolomite frequently occurs as crystals showing a pronounced saddle shape; prehnite, calcite, mimetite, and adamite are good examples of minerals occurring frequently as aggregates of crystals exhibiting curved surfaces, even forming spheres!

Curved crystal-face surfaces are really not uncommon at all, but true *twisted* crystals are quite unique and scarce, one of the most intriguing mineral oddities.

The most notable examples are the smoky- or clear-quartz crystals from the Grimsel, Goschener Tal, and Val Guiv localities in the lofty Central Aar Massif region of Switzerland. The alpine clefts and vugs of Switzerland have produced excellent quartz-crystal specimens since the early 1700s, many of them among the finest in the world. Twisted crystals of both smoky and clear quartz are always found associated with crystals of normal habit. More twisted specimens are encountered of smoky quartz than of clear, and the cause of the twisting remains somewhat of a mystery. The Swiss-German word for this phenomenon is *Gwinden.* A twisted crystal from the Swiss localities looks very much as if someone grasped the crystal when it was soft and gave it a good twist, thereby giving the crystal a propellerlike appearance when viewed from one angle. The amount of pitch, or offset, can be either very slight or marked and differs greatly between individual specimens.

Actually, these exotic quartz crystals are morphological oddities composed of multiple crystallizations, one beside the other, and each of the lamellae offsets a small angle so as to produce a modest spiral, corkscrew, or warped effect. They are complex multiple intergrowths or aggregates of numerous individual crystals—but this is seldom apparent in viewing a specimen.

From the standpoint of crystallography, these multiple crystallizations usually take place along one of the a-axes of the hexagonal crystal prism, producing the elongated effect, and they are also distorted around this same crystallographic axis. The offset of one merged crystal from the other is so slight that it produces amazingly smooth, lustrous, curved surfaces on the twisted faces. If it is impossible to detect the individual crystals composing the specimen, it is called a *closed crystal.* If, however, individuals of the composite may be observed—like teeth marks along the edge of the crystal—the specimen is called an *open crystal.*

Another approach to these unique specimens is to think of them as a composite of numerous crystals occurring as a stacked and compressed aggregate. They are usually found oriented with the crystallographic c-axis horizontal to the rock matrix. Interestingly enough, analysis indicates that the twisted phenomenon is characteristic of the aggregate as a whole, not of the individual components.

Twisted quartz crystals have been found at other localities throughout the world—mainly in Colorado, California, Italy, and the Ural Mountains. However, they do not approach either the abundance or the quality of the Swiss specimens.

Good specimens of either smoky- or clear-quartz twisted crystals can vary in size from around 1" up to 10". Many of the finest Swiss specimens are virtually, if not completely, flawless, a factor which adds a dramatic dimension to the specimens. Fine specimens are considered very choice collector's items.

Figure 3-1. A strongly *twisted* closed crystal of smoky quartz from near Fellital in the Aar Massif region of Switzerland. These mineralogical oddities are fine collector's items, and good examples are difficult to obtain.

Scepter Crystals

Another remarkably attractive type of unusual crystallization in quartz is *scepterism.* Although the finest examples of scepter crystals are found in quartz, other minerals such as calcite can show this phenomenon.

As you may gather from the name, these are crystals whose growth has produced a form resembling a monarch's scepter. They are usually characterized by an *overgrowth* of one crystal on top of another. The most common form is a slender prism of quartz capped with what is obviously a second crystal, usually in a doubly terminated form. Frequently the basic quartz prism—the slender supportive host—is one color and the scepter capping is another! For example, the prism might be somewhat milky, capped with a smoky scepter; a smoky prism, with a citrine-colored scepter; or a clear prism, with a fine amethyst scepter!

Most scepter crystals are definitely overgrowths initiated after the original crystal has been arrested, and they occur under different conditions, temperatures, and chemical content of the crystal-forming material. This accounts for the different-colored scepter cappings, which indicate a secondary stage of growth development. It is interesting to note that the scepter portion of such a crystal is usually much clearer than the supporting crystal prism, probably indicating a slower, more stable crystallization condition.

Specimens have been found at Llano, Texas in which the scepter cap can actually be lifted off! In this case the scepter failed to seal itself to the original supporting crystal but continued perfect scepter growth.

The reasons for scepterism are not fully understood. Why a reinitiated growth stage of a quartz crystal should favor development on the termination, or tip, of the crystal as opposed to overall facial growth is a mystery. The hesitant growth of quartz crystals is commonly observed in crystals showing from one to a dozen or more phantoms, in which the normal crystal form is retained. The secondary growth of a scepter usually orients the bulbous scepter faces parallel to the original prism faces of the host crystal. This shows that some atomic influence must still be manifested by the original crystal, holding the second growth in line. However, the scepter may sometimes be slightly tipped, yet it maintains its basic plane alignment with the original crystal. Many questions about observable crystallized-mineral phenomena remain unanswered.

In addition to the obvious capping overgrowths, there is another type of scepter in which absolutely no discernible secondary-growth pattern at all is exhibited. The crystal may be of even clarity or opaqueness throughout and yet show a perfect, strong scepter either at the end of the original prism or somewhere near the base! This, too, may be the product of arrested growth, but with no change in the chemical content of the silica solutions involved. Fine specimens of this type, large and small, have come from many of the Swiss localities as well as Brazil and a few from Arkansas and California.

Perhaps the most remarkable examples are the clusters of tiny, strongly sceptered, water-clear quartz crystals found inside agate nodules from a particular bed, or stratum, near Dugway, Tooele County, Utah. The perfection and brilliance of these little crystals must be seen to be believed. They represent secondary crystallizations inside the hollow cavities of bluish chalcedony-lined geodes. Along with the remarkable scepters are found outstanding examples of skeletal- and cavernous-type quartz crystals.

Wherever quartz crystals are found, scepters may occur. Dozens of localities have produced good specimens. They are always found in direct association

Figure 3-2. *Scepters* are among the most unique of all natural-crystal forms. These quartz crystals are terminated in scepters of smoky amethyst, and a peculiar pinkish, transparent quartz. Usually the scepter is a different color than the host-crystal prism. From left to right these specimens are from Mexico, Korea, and Texas.

with crystals of perfectly normal habit, although some localities such as Llano, Texas and Hallelujah Junction, California may contain a comparatively high percentage of scepter-oriented crystals. Although scepter quartz crystals are relatively uncommon, they are more abundant than twisted crystals. Spectacular scepter specimens have come from many localities in Switzerland and Austria. Brazil, Korea, France, Japan, and Mexico have also produced choice specimens.

At least four different localities in the United States have produced a few very unusual and attractive quartz-crystal prisms with scepter overgrowths of either amethystine or smoky quartz on both ends of the crystals. These very scarce, doubly terminated scepter specimens look very much like dumbbells. They have come from the well-known Hallelujah Junction locality and nearby Doyle in Lassen County, California. There was also a very productive find east of Bishop, Inyo County, California, as well as at the former Pohndorf Amethyst mine, southwest of Butte in Jefferson County, Montana. Similar specimens have been found in Brazil.

Staurolite

Would you believe that a perfectly natural crystal can look unreal and man-made? This is the result of the appealing cruciform habit of the mineral *staurolite.*

Staurolite is a dark brown, iron-aluminum-silicate mineral found at many localities throughout the world in various metamorphosed schist-type rocks. It is usually opaque. Unaltered or nonweathered specimens have a hardness of 7 on a scale of 10; consequently, these little crystals are quite durable. However, staurolite is also found in a severely altered talclike material that is very soft. Staurolite crystals are usually small, varying from ½″ up to 1¼″, but crude crystals up to 2′ and 3′ have been found.

What makes staurolite particularly unusual is that in an overwhelming number of localities where the mineral is found it occurs almost exclusively in a well-crystallized state. Twinned crystals of the mineral are exceptionally common, if not the rule, at some localities.

Staurolite crystallizes in a single orthorhombic pattern, frequently as oblique penetrating or intertwined twins at 60° to each other but more frequently at right angles, forming almost perfect Maltese- and Roman-style crosses!

These interesting natural crosses have been used as amulets, religious symbols, and good-luck charms in both Europe and the United States. The mineral name is derived from the Greek word for cross, *stauro*, and specimens have been sold commercially for many years as "fairy stones," "fairy crosses," or "lucky crosses."

There are numerous localities for this mineral. Of particular note are several localities in Patrick, Henry, and Franklin Counties in Virginia. Here the little crosses are found loose in the topsoil by the hundreds. Some erode out of ledges of schist-type rock and have been a local source of income here since the late 1890s. As a matter of fact, Virginia created a Fairystone State Park; however, few staurolites are found in the park area itself. There are several good localities in adjoining North Carolina, in Georgia, and in a number of sites in Massachusetts, Connecticut, Maine, New Hampshire, Vermont, and Minnesota. A few somewhat crude specimens have come from Idaho and Montana. In the Southwest, however, there is only one truly spectacular locality for staurolite specimens. This is the beautiful mountainous area between San Lorenzo Pueblo (Picuris) and Taos Pueblo, Taos County, New Mexico. This large area has furnished thousands of fine, almost perfect staurolite crosses. They are found loose in the numerous washes or in the topsoil, as well as in the host-rock outcroppings along ridges and canyon walls.

Mention should be made of one very special foreign locality for staurolite, attractive specimens from which are often seen in museum collections. This is the locality of Ticino and Pizzo Forno, Switzerland. Here a type of paragonite schist is found, composed essentially of silvery-colored mica. This rock contains slender crystals of sky-blue kyanite associated with elongated, single-prismatic, brownish staurolite crystals. A few of these lustrous staurolite specimens have been found with enough transparent sections to yield small faceted stones of a pleasing fawn color.

Along with this apparent abundance of natural specimens goes a particular aspect of fraud! Many staurolite twins are not particularly distinct and sharp when found. Although the basic crystal form of the cross is well defined, many are "improved" for sales purposes by filing and sanding. These hand-modified crystals are then soaked in glycerine, linseed oil, or motor oil to lend a dark-brown appearance. They will frequently show file marks. Further, regardless of the abundance of natural specimens, many so-called fairy stones or fairy crosses are actually completely handmade by carving, sawing, and filing cross-like forms from a soft brownish-to-reddish talcose schist or a claylike material. This material bears no relation whatsoever to true staurolite. Such faked crosses are soaked in oil, buffed, and affixed with a silver or gold mounting and are sold as the genuine thing to thousands of unsuspecting tourists. This has been true particularly in the southeastern states.

Roses, Rosettes, and Spheres

In descriptive mineralogy (those aspects of a mineral specimen that may be readily observed or ascertained by simple tests) there are a number of categories. These are called the physical properties of minerals and involve color, streak, luster, optics, index of refraction, hardness, specific gravity, cleavage and parting, fracture, tenacity, electrical properties, magnetism, luminescence and radioactivity, and structure. A crystal is only one of several categories under structure but is often the most desirable from an identification standpoint. In addition to crystals, a mineral may occur in any of a dozen or more different physical forms. There are several descriptive terms in mineralogy that specify the fundamental form, shape, or physical appearance of a given specimen. An *acicular* specimen is found in long, thin, needlelike crystals; a *druse* is a coating of small crystals on a matrix; a *foliated* specimen is composed of thin, separable leaves or sheets; and an *arborescent* specimen suggests a treelike branching form. There are many other descriptive terms in common use, but these are typical.

Three particularly interesting physical forms found in certain minerals from selected localities are *roses*, *rosettes*, and *spheres*—or spheroids. Since these three descriptive categories often prompt misconceptions among beginning collectors and are always puzzling oddities to the novice, more details are in order.

Essentially, all three terms refer to clusters or aggregates of crystals whose external form mimics, in various degrees, a roselike, rosettelike, or spherical shape. Such forms are absolutely not fossil roses by any stretch of the imagination! The terms "rose" and "rosette" are virtually interchangeable, but "rose" is used for a spherelike cluster, and "rosette" for a more flattened cluster.

Figure 3-3. *Fairy stones*, or *fairy crosses*, are intriguing natural occurrences of the mineral staurolite in cruciform penetration twins. They occur in both Roman and Maltese styles, as seen in the bottom specimens from New Mexico. The specimen on the upper left is from Georgia; upper right, a fake, fashioned for the tourist trade from a claylike material.

The minerals barite and gypsum, found in many localities throughout the world, often occur in more or less spherical clusters of well-developed crystals. These clusters may easily resemble a rose in full bloom. Oklahoma, California, Mexico, Canada, and Egypt have been sources of good specimens. In size, roses may vary from around 1″ to huge specimens from Mexico measuring 1½′ in diameter. The barite roses of a deep red-brown or cinnamon color from Norman, Oklahoma are classic examples, and fine specimens are abundant.

Specimens may occur as single roses or as large aggregate clusters of dozens of roses. Sometimes the latter are marketed by decorators under the name "rose rock," an easily misinterpreted term.

There are a few minerals, especially azurite, that crystallize at selected localities in beautiful ball-like rosette forms. Bisbee, Jerome, and Ajo, Arizona have furnished dazzling small specimens of exceptional quality. Similarly, the old locality of Chessy, France has yielded excellent rosettes of azurite, some partially altering to malachite-variety "chessylite."

Switzerland has produced many stunning hematite roses. These are brilliant clusters of bright, metallic bladelike crystals of hematite in a circular roselike form, often laced with lustrous crystals of red to sherry-colored rutile.

Several minerals may occur in marblelike spheres, particularly mimetite, smithsonite, adamite, marcasite, and prehnite. They are always among the most curious of specimens. Actually, such forms are simply an aggregate of many crystals growing in concert or multiplying the phenomenon of a normal curved crystal face.

Figure 3-4. *Roses* are one of the most fascinating of all mineral forms. They are aggregates of several crystals growing in a clustered configuration. Many minerals may occur in this form. At the top left is a barite rose from Oklahoma; the other three specimens are gypsum roses from different localities in southern California.

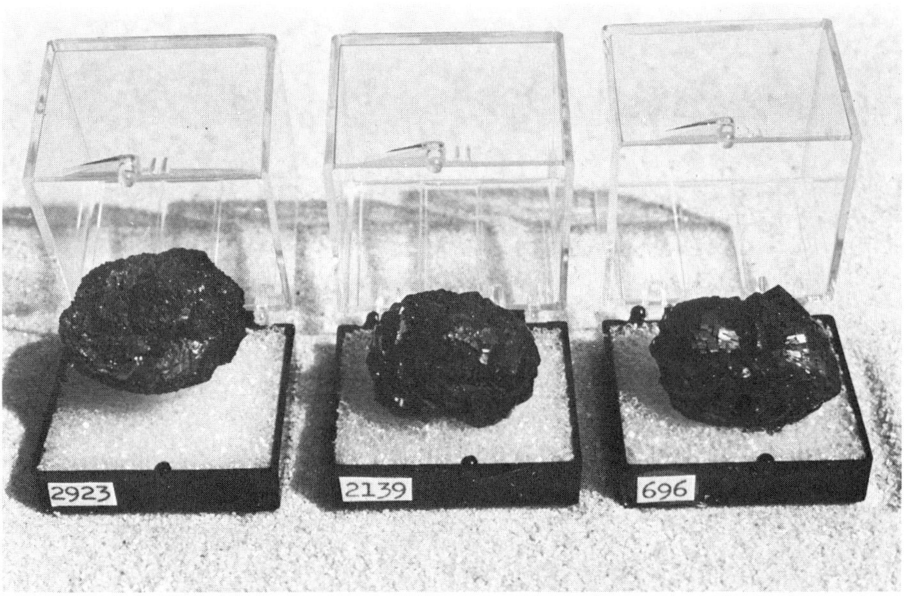

Figure 3-5. Deep blue azurite is one of the most appealing minerals, and it occurs at many localities in fine *rosettes*. These three specimens are displayed in 1¼" plastic boxes and from left to right originate in Jerome, Arizona; Zacatecas, Mexico; and Bisbee, Arizona.

Figure 3-6. A ball-like shape is one physical form in which a number of minerals may occur. These *spherical aggregates* of many individual crystals grow in unison and form a spherical or oblate shape. They can be exceptionally beautiful specimens. From top to bottom, left to right these are: adamite, Mexico; prehnite, New Jersey; calcite, South West Africa; smoky quartz, Arizona; goethite pseudomorph after pyrite, Austria; wavellite, Bolivia; and azurite-malachite-variety chessylite, France.

Figure 3-7. A fine "iron rose" of hematite associated with adularia crystals. This specimen is from the famed Alpine locality La Fibbia, near Gotthard Pass in Switzerland. This brilliant metallic-appearing aggregate of hematite crystals measures 1¾" × 1½".

Bow Ties

One of the oddest shapes certain minerals can assume is that of an hourglass, or, more commonly, *bow tie*. Obviously, this is a metaphoric term alluding to the resemblance of this type of specimen to a bow tie. Specimens in museums, mineral shows, and private collections always prove fascinating since it is difficult to believe that they are perfectly natural!

The mineral stilbite, in particular, is most often associated with this form, although a limited number of localities have produced fine bow-tie specimens of strontianite, calcite, and quartz.

Actually, this bow-tie appearance is simply an unusual crystallization form of a doubly terminated sheathlike cluster of many crystals. Upon close examination it looks like a bundle of slender crystals that have been pinched together in the middle and rounded at the ends! The degree of "bow-tieness" may be minimal, showing only a slight flanging, or it may be radical, with the tips of the bow ties almost touching each other.

Outstanding specimens of stilbite bow ties have come from the famed traprock quarries in New Jersey. Specimens have been found with pearly-tan or pinkish bow-tie clusters up to 5" in length. These unusual specimens are associated with other normal crystals of stilbite, chabazite, calcite, prehnite, and apophylite.

Wherever the various members of the zeolite mineral family and their related associates are found, there is the possibilty of bow-tie crystallization. However, bow ties are never the dominant crystallization form at a given locality, and several localities that produce very fine collector-quality specimens of stilbite have none of this form.

Figure 3-8. Several minerals occur in the imaginative form of a *bow tie.* These forms are actually a composite of doubly terminated sheathlike clusters of many individual crystals. Stilbite bow ties have occurred in classic examples from New Jersey, Iceland, and India. This specimen is from the Deccan Traps formations near Poona, India.

In addition to New Jersey and its environs, other superb specimens of this type have come from the zeolite-producing localities in the Bombay-Poona region of India. Here very large, thick, lustrous specimens of a creamy-pink-to-peach color have been found, and they grace the cabinets of many fine mineral collections. Smaller specimens of a rust-to-orangish-brown color have been found for many years in the rocky cliffs surrounding the Bay of Fundy, Nova Scotia, Canada. Recently, spectacular specimens of a cinnamon-to-deep-tangerine hue have come from New South Wales, Australia.

Crystals of Unusual Size

In reading the above heading, you are probably thinking of very large specimens. As mentioned earlier, huge crystals do occur; however, there is another much more sophisticated and fascinating aspect of size that has special appeal to the mineralogist and the avid mineral collector.

An appreciation of the usual versus the unusual in individual crystal sizes of certain minerals comes only with study and familiarity with some of the more challenging nuances of mineral collecting. There are hundreds of examples that could be discussed that would be familiar to the museum curator or the advanced collector; however, one example should suffice to make the point.

Some minerals, both rare and common, may frequently occur in a wide variety of sizes, and this property of a mineral is a well-established physical characteristic. Quartz, for example, occurs in superb microscopic crystals of great perfection as well as in perfect crystals weighing 400 or 500 pounds! The same may be said of fluorite or pyrite, with crystals ranging from the microscopic to single crystals of 8″ or 10″ on a face.

The mineral mimetite is a good example of size and oddity. Mimetite is an attractive secondary mineral often found well crystallized in the oxidized zones of lead-ore deposits. It is a complicated chloride-phosphate-arsenate of lead and shares a virtually identical chemical formula with its sister mineral, pyromorphite. It is found at many different localities throughout the world, in a wide range of colors, and almost always as very small crystals up to ½″ in length. It also occurs in clusters of small spherical aggregates, particularly from various Mexican localities. Fine examples are choice collector's items. The mineral is somewhat uncommon, but assuredly not rare; however, large crystals are indeed uncommon!

In recent years a pocket of superlative water-clear, yellowish crystals on matrix was removed from one of the mines at the Tsumeb mining complex in South West Africa. This locality had long provided excellent mimetite crystals, but not like this find, since the crystals ranged up to ½″ wide by 1″ long. That may not sound big, but it surely is for mimetite! Fine miniature specimens sold for well over $1000—so imagine what the going price would be for cabinet-size specimens!

Another beguiling example of huge mimetite crystals is a find made in the early 1960s at the Broken Hill mine, Broken Hill, Northern Rhodesia (currently Kabwe, Zambia). A very few truly remarkable mimetite specimens were removed as floaters within a claylike mass of material in a pocket. The specimens are lustrous plumose aggregates of hundreds of yellowish-to-taffy-colored mimetite crystals clustered together in thick sprays that look like mushrooms. The "mushrooms" vary from 1″ to 1¾″ in length, and again, that may not sound big, but it is! They are truly odd and very choice.

Fulgerites

This may seem like an unusual place to include lightning, but it strikes me as quite appropriate.

A *fulgerite* is a branching, glassy, tubelike formation or vitreous crust formed on bare rocks by lightning. Fulgerites are common in high-silica-content sands in deserts and in other areas subjected to lightning storms, such as the sand dunes of the Kalahari and Sahara Deserts, many dune areas in the western United States, and commercial sand-pit operations along the eastern seaboard.

When a powerful bolt of lightning strikes a sand dune, it branches erratically through the sand. As it courses its way downward, it melts and fuses the sand granules into branching tubelike forms that resemble the roots of a tree or shrub.

Fulgerite color is determined by the chemical content of the particular sand involved; most fulgerites are a grayish blue or black. Actually, a fulgerite is natural silica glass. The tubes are rough, sometimes spiny, and are covered with adhering sand on the exterior, while the inside is hollow, smooth, and glassy. Fulgerites are usually about ½″ to 1″ in diameter, but tubes up to 4″ in width have been found. A fulgerite labyrinth may extend for 30′ to 40′ into a sand dune. As the bolt is discharged, its path is deflected by rocks or pebbles, which perhaps causes the remarkable branchlike structure of this truly odd formation.

Fulgerites may have a series of from one to four corrugated ridges on their surface, often oriented in a spiral pattern. The spiral is always right-handed!

I have seen fulgerites sold as "lightning silicified-root systems," the story being that the bolt struck a shrub and instantly turned its roots to glass. There is no data to support this, and fulgerites are very common in some desert areas

Figure 3-9. When lightning strikes the earth's surface in a sandy area, it may course its way downward, fusing the sand grains into strange branching, tubelike formations. These are called *fulgerites* and are mineralogical evidence of an atmospheric discharge!

completely devoid of any major plant life.

Lightning also frequently strikes high mountain peaks, and there may be a fusible element in the rocks involved. Such strikes may leave a glazed crusty area, which is also called a fulgerite.

Extraterrestrial Specimens

Since fulgerites are caused by quasi-earth-connected phenomena, perhaps we can carry this subject a bit further out! A few capsulated paragraphs regarding extraterrestrial specimens might be in order, since they are assuredly odd and very uncommon. The term refers to materials found on the surface of our planet that did not originate here. They represent true galactic visitors from the vast voids of outer space.

The "shooting star" seen in the evening sky is a meteor, and if it lands on the earth's surface and is found, the specimen is called a *meteorite*. There are many different kinds of meteorites, and they are classified in terms of their composition. Those that are composed essentially of iron and nickel are usually referred to as "iron meteorites" or "meteoric iron" and are called *siderites*.

The nickel content is usually between 6 and 10 percent but can be higher. When they are cut and polished, many iron meteorites look like steel. Often this type of meteorite will show an odd pattern of triangular intersecting lines, called Widmanstatten figures, when the polished surface is etched with acid. This suggests a partial octahedral structure, and consequently this type of meteorite is called an *octahedrite*.

The term *siderolite* is used for another major grouping of what are basically stony iron meteorites. These look somewhat like a solid sponge, in which the

Figure 3-10. This is a group of *meteorites* from outer space. The polished specimen and the large meteorite below it are from the Meteor Crater, Arizona and are called *siderites*. They are composed of iron and nickel. The middle specimen from Plainview, Texas, mainly composed of stony material with minor flecks of metal, is called an *aerolite*. The small specimen is a siderite from Henbury, Australia.

sponge itself is iron, and the holes are filled in with various silicate minerals, composed of easily defined proportions of both metallic and stony materials.

Meteorites that are essentially composed of stony materials with very minor flecks or particles of iron are called *aerolites*. Most meteorites are very heavy, but a true aerolite can be amazingly lightweight as well as very difficult to identify.

Don't bother going out to hunt meteorites—the chance of finding one is infinitesimally remote! A mere fraction of the meteors we see flashing across the sky ever reaches the earth's surface. The reason you see a meteor is because it is burning, due to the friction of air against the hurtling mass. Most burn up in the atmosphere, and about three-quarters of the small number that reach the earth's surface fall into the ocean.

The stereotyped comic-strip concept of a meteor involves a huge, white-hot, cannonball-shaped object zooming through space like a great fireball. Meteors may be a single mass, but they certainly are not round. At best they might be called very irregularly shaped spheroids, roughly pyramidal or conical in shape or deeply pitted ovoids. Meteors may be composed of several individual masses hurtling through space as a unit or even hundreds in a swarm.

There are a few areas where a swarm, or meteoric mass, has struck the earth's surface, creating a huge impact crater or group of craters and scattering meteoric iron over a wide area. At such a locality you might find a specimen. Many hundreds of specimens have come from around Plainview, Texas; Henbury, Australia; and Arizona's famed Meteor Crater, near Winslow. A large meteoric mass, such as the one that formed the impact crater in Arizona, probably struck the surface with a velocity of about 20,000 miles per hour. Meteorite specimens, varying from the size of a pea up to several tons, have been found at about fifteen hundred localities. Nevertheless, they remain extremely rare.

The next *possible* category of extraterrestrial specimens is *tektites*, a somewhat controversial and certainly curious material. Taking into account the fact that many scientists disagree on tektites and their meteoric origin, following is a brief outline of the subject.

Tektites come in a wide variety of shapes. They are small and may be irregularly spherical, lens- or button-shaped, pearlike, dumbbell-like, and even spindle-shaped. They are actually droplets of a special type of siliceous glass. The glass has a very high melting point and a complicated and odd composition for such material. In hardness tektites vary from 5 to 6 on a scale of 10. Tektites appear black on the surface, but if they are broken or held to a strong light, they may be of a translucent brown hue, light or dark green, or smoky. The name comes from the Greek word *tektos*, which means "molten."

Tektites are almost always mistaken for obsidian. The similarity is, indeed, exceptional, particularly when tektites are compared with one of the spherical obsidian nodules called "Apache tears" or even with a small waterworn pebble of obsidian. Obsidian is of volcanic origin; however, tektites are found abundantly in some localities where there is absolutely no indication of volcanic activity. The chemical composition of the two materials also differs considerably.

Figure 3-11. *Tektites* are among the most curious of all materials and are of extraterrestrial origin. They are actually a special type of glass. These specimens are from Cambodia and Indonesia. They are easily mistaken for obsidian, and the similarity can be seen in the two upper-right-hand specimens—a waterworn pebble of obsidian and an obsidian nodule called an Apache tear.

Tektites are believed to be of cosmic or interplanetary origin. Some believe that they are the result of huge meteorites striking our moon and "splashing" rock materials into space. This material in turn is attracted by the earth's gravity, assumes a gradually decreasing orbit around the earth, and eventually strikes its surface. These fascinating little glass objects often show sculptured surfaces featuring very marked pitting and grooving. These can well be evidences of what appears to be ablation, and these patterns make a striking case for a journey through space. Researchers have even backtracked tektites from their earthly localities and computed the prospective craters on the moon which would fit the orbital disintegration pattern correlating with the site on earth!

There are several major areas where tektites have been found in varying degrees of abundance for many years. Some localities have produced many thousands of specimens. Often these have been given local names, but they are still the same type of material, sharing the same group of common denominators. Since the early 1900s "moldavites" have been found around Moldavia, Czechoslovakia. Hundreds of fine specimens have come from Cambodia and Thailand and are called "indochinites," while "billitonites" come from Billiton Island in Indonesia. The Philippine Islands produce "risalites," and a vast area in southern Australia has furnished thousands of "australites," while Tasmania is the source of "Darwin glass." In the United States a locality involving a portion of east-central Texas and extending into Georgia contains infrequent "bediasites." The unusual scatter pattern, or *strewing*, of tektites at their localities is another factor that lends substance to the theory that swarms of these cosmic materials were sown across the earth's surface at various times.

Chapter 4

THE FASCINATING WORLD OF INCLUSIONS

What are inclusions in mineralogy? Bubbles that dance back and forth within crystals; individual crystals showing several colors; ghostly phantoms of different crystals inside each other; different minerals seemingly suspended or frozen within another crystal; agates with treelike or flowerlike growths; rocks that rattle or splash when shaken—some even filled with oil!

As surprising as it may sound, none of the foregoing are particularly rare in the field of mineralogy. They are fascinating examples of the chemistry that has existed for millions of years in nature's great underground laboratory—the crust of the planet Earth. The following information is an introduction to some of the more common and interesting types of inclusion phenomena.

Figure 4-1. Suspended like jewels frozen in ice, brilliant brassy pyritohedrons of pyrite float in flawless clear quartz. These spectacular *inclusions* of pyrite in quartz were found in the early forties near Diamantina, Minas Gerais, Brazil.

Unquestionably, some of the world's most spectacular and provocative crystalized-mineral specimens are those showing the natural phenomena of *inclusions*. In essence, this means that one mineral encloses, or displays, one or more other minerals *inside* its crystals.

It is very common for crystallized minerals to occur in *combinations*, *intergrowths*, or *aggregates* of two or more minerals on the same specimen. As a matter of fact, many of the crystals viewed in major museum or collector displays will show combinations of two or more minerals on the specimens. Usually these are cataloged or referred to by naming the more important mineral of the combination first. For example, a specimen might be labeled "calcite, sphalerite, and galena," indicating that in this particular case the calcite is the most prominent mineral on the specimen, the sphalerite second, and the galena third. This is a typical and very common example of a combination of crystallized minerals occurring together. The rock, or solid-mineral material upon which the crystals have grown, is called the *matrix*.

Keeping in mind that combinations or intergrowths of crystallized minerals are very common, another descriptive phenomenon is often encountered, that of *encrustation*. This term refers to a special format of mineral combination in which one mineral coats or covers another. A cube of fluorite may be completely covered with siderite, danburite with calcite, or hematite with quartz. The covering is usually caused by a second stage of mineral deposition in a cavity or pocket in an ore body. Mineral-laden solutions or gases enter an area that is already well crystallized and deposit their substance as a crystallized covering, or encrustation, over the first.

Inclusions in Crystals

From one standpoint any major ore body can be considered to contain nothing *but* inclusions: i.e., a gigantic compound constituting a complex admixture of many minerals in the rock of the ore. Quite true, but the types of inclusions sought by mineralogists, museum curators, and advanced collectors are those in which at least one mineral can clearly be seen to be trapped or suspended within another. Obviously, in this type of specimen the host mineral must be translucent or transparent. The combinations are virtually unlimited, often spectacular, and among nature's most remarkable phenomena.

Since silica is the most common mineral element in the earth's crust, it follows that quartz is the most abundant crystallized mineral—water excepted. Quartz crystallizes in virtually all types of ore bodies and consequently carries the highest potential for involving other minerals as visible inclusions. Also, quartz is commonly quite clear, and its crystallization with other minerals is legion—there are numerous combinations from hundreds of localities.

Perhaps the most stunning examples of inclusions in quartz are seen in the fantastic crystallized specimens from Brazil. Here very large quartz crystals have been found containing lustrous radiating bursts of acicular ruby-red or golden crystals of rutile. These are often associated with plumes of greenish chlorite, swirls and plumes of white and pink clay minerals, floating bladelike rosettes of hematite, and even perfect pyritohedrons of pyrite suspended as if frozen in ice. The beauty and brilliance of these specimens are unsurpassed. Many very handsome inclusion specimens of various minerals in quartz have also come from Switzerland, Rhodesia, Portugal, Japan, Mexico, California, and Arkansas.

A small, isolated locality south of Searchlight, Clark County, Nevada has produced thumb-sized clear-quartz crystals containing tiny, perfect, transparent red garnets as inclusions. Various localities in California and in Baja California, Mexico have produced quartz crystals literally filled with black tourmaline, various fibrous amphibole minerals, epidote, chlorite, and even some native gold.

A particularly notable inclusion find was made several years ago in a quartz-lined pocket in a mine at Messina, Northern Transvaal, South Africa. Here translucent quartz crystals enclosed inclusions of brilliant sky-blue fibrous crystals of papagoite, as well as occasional tiny spherules of red hematite! Even though quartz is a very common crystallized mineral, inclusions of this type produced specimens of incomparable beauty and interest.

Calcite is another mineral containing an abundance of inclusions. Here, as in quartz, we are dealing with a very common mineral encountered under a variety of ore-body and deposit conditions. Calcite is often very clear and is commonly found associated with dozens of other minerals; consequently, it has a high potential for cocrystallizing with them and enclosing them as inclusions. Perhaps the most spectacular are the inclusions of brilliant metallic native copper in calcite crystals from the Keweenaw Peninsula in Michigan. These are true American classic specimens. Similar specimens have also come from copper ore bodies in Nevada. Fascinating large sprays of brilliant arborescent native copper have been found in transparent sheets of gypsum-variety selenite from the Mission mine near Tucson, Arizona. They do not even look real! Mexico has yielded a host of localities producing exquisite specimens of calcite with exceptionally showy inclusions of vanadinite, wulfenite, malachite, pyrite, aurichalcite, and chalcopyrite. When it comes to sheer breathtaking beauty, many of Mexico's suites of inclusion specimens would be difficult to top.

Other fairly common minerals that frequently contain mineral inclusions are fluorite, celestite, barite, and hemimorphite. As a matter of fact, almost any mineral occurring as a translucent or transparent crystal may have visible inclusions. Many diamonds have tiny inclusions of black carbon, a factor which may decrease the value of the stone.

Phantoms and Zoning

A fascinating dimension within the inclusion field is that of crystal specimens showing *phantoms*. These shadowy ghostlike forms inside a crystal show former stages of the crystal's growth. As the crystal grows, various other mineral agents may be introduced in the process, adding a strong color line or faint, hazy shadow at that particular stage of development. Quartz crystals showing as many as twenty distinct phantoms stacked on top of one another like caps have been found in Brazilian specimens. Phantoms usually favor certain crystal faces, often leaving unphantomed faces, thus producing a windowlike effect that allows the viewer to look in on the former growth stages of the crystal.

Phantoms are not particularly rare, and most deposits that yield quartz crystals have specimens showing some degree of the phantom phenomenon.

A truly unique locality in the mining district surrounding Manhattan, Nye County, Nevada has produced finger-sized, clear-quartz crystals containing perfect phantoms composed of silvery metallic stibnite. The phantoms look much like miniature haystacks inside the quartz crystals. This same locality has even produced small doubly terminated quartz crystals containing doubly terminated stibnite phantoms! Specimens like this lend a superdimension of both value and interest to the most common of minerals, quartz.

Very attractive calcite crystals showing perfect phantom microcrystals of pyrite have come from Wanlockhead, Dunfriesshire, Scotland. Similarly, the famed Tri-State District of the United States has produced hundreds of outstanding brilliant yellow-to-golden calcite crystals that are well phantomed by pyrite, marcasite, or sphalerite, clearly showing a former state of the crystal's growth. In 1974 a pocket was opened in the Sweetwater mine of Missouri's St. Joseph Mining District that furnished a number of magnificent large, brilliant scalenohedrons of yellow-green calcite containing amazingly distinct phantoms. Various localities in England as well as at Tsumeb, South West Africa have produced large, water-clear calcite crystals enclosing perfect phantoms etched in deep red hematite. Needless to say, these are spectacular specimens.

Another interesting aspect of phantoms depends on crystal *habit*. Many minerals occur in well-established but different crystal habits, or forms, while others crystallize in only one form, or, if you prefer, one shape. Calcite is the most versatile of all minerals with over twelve distinct habits, some of which are very complex. Both Scotland and the Tri-State District have produced calcite crystals showing two distinct types of crystallization in the same crystal: for example, a scalenohedral crystal enclosing a perfect phantom of a "nailhead"-habit crystal, or vice versa.

The growth phenomenon of phantoms is often referred to as *zoning*. Outstanding examples of zoning occur in the mineral fluorite. The various mines in the district around Rosiclare and Cave-in-Rock, Hardin County, Illinois have produced some exceptionally beautiful zoned fluorite crystals. The specimens are quite transparent with exceptionally well-delineated zones of yellow, golden

Figure 4-2. Quartz, fluorite, calcite, and selenite often show *phantom* inclusions composed of other minerals deposited on crystal faces during various stages of growth. These quartz crystals contain phantoms of chlorite and clay minerals. The crystal on the left is from Grass Valley, California; the other two, from Minas Gerais, Brazil.

yellow, and shades of purple. Fine specimens create the impression of looking through a ghostly series of cubes, one inside the other and each of a different color.

Another mineral famous for spectacular zoning is the gemstone tourmaline. Tourmaline specimens from many localities, particularly Madagascar (Malagasy Republic); Mozambique; Minas Gerais, Brazil; Maine; and San Diego County, California, have exhibited crystals showing extraordinary color changes, with either a horizontal or longitudinal orientation, within a single crystal. Two colors, such as pink and green or pink and blue, are common, and crystals clearly showing four colors are not uncommon! The color-differential zone can either be very sharp or show a very subtle progressive blending of colors throughout the length of the crystal.

Many classic specimens from Mozambique and Madagascar show a watermelon effect. When the specimen is cut in thin slices at a right angle to the crystal prism, it will often show concentric bands of different colors.

A number of superlative specimens from various localities in Minas Gerais, Brazil, as well as from the mines surrounding Pala and Mesa Grande, San Diego County, California, may show astonishing zoning throughout the length of the specimen. For example, a single crystal may have a ruby termination, phasing into pink, green, and dark purple at the matrix. Various color combinations are highly characteristic of individual mines and even individual pockets within the pegmatites in which the mines are located.

Mineral Color and Inclusions

An intriguing but often overlooked aspect of inclusions involves mineral *color*. A given mineral may be characterized by a common color, but the same mineral may occur in an incredible variety of colors as well! The colors are due to the chemical presence, or inclusion, of minute portions of other elements while the major mineral is crystallizing. For example, most people associate sulfur with just one color—yellow. However, depending on the chemistry associated with its deposition at different localities, sulfur can vary from a brilliant light lemon color through various shades of deep yellow, green, canary, orange, and even brown.

Another example that is a top favorite among many mineral collectors is the lead mineral wulfenite. Wulfenite occurs attractively crystallized at many localities throughout the world and can be colorless, black, tan, brown, yellow, orange, caramel, tangerine, scarlet, ruby red, lime, and even light blue!

For a subtle rainbow of potential colors few minerals can rival smithsonite. Basically a carbonate of zinc, smithsonite can occur in either crystallized or botryoidal specimens in an astonishing array of colors, varying from pure white and mauve through the grays, yellow, orange, green, blue, peach, rose, lavender, and lilac! Smithsonite is a particularly attractive and interesting mineral and is always a favorite among discriminating collectors.

Liquid Inclusions

The fascinating subject of *liquid* inclusions in minerals breaks down into two fundamental categories: inside crystallized minerals and inside nodules and geodes. In both cases astonishing specimens have been found. With perhaps one unique exception—the bitumen-filled geodes of Niota, Illinois—such specimens have logical explanations. Water-filled geodes, called enhydros, and oil-filled geodes are discussed in the following chapter.

Liquid inclusions in crystallized minerals occur more frequently than you might imagine. Fluid inclusions (usually grouped together erroneously as "bubbles") occur in quartz crystals from many localities; they are frequently found in gypsum-variety selenite, and infrequently in fluorite, rarely in calcite, and even more rarely in emeralds.

The liquid involved is normally water, although liquid carbon dioxide has been reported. In the case of the minerals mentioned (quartz, fluorite, selenite, and calcite), water is often present and is a direct or indirect element of the original crystal formation and/or composition. During crystal growth water is apparently trapped in cavity areas within the crystal.

Most of the time an internal cavity within a crystal is either empty (except for the air or other gas it may contain) or is completely filled with liquid. Sometimes there is an air pocket within the cavity that produces a discernible, movable bubble. As such specimens are examined closely, the imprisoned bubble may be seen to move about within the cavity. Sometimes the movement is almost imperceptible, due to the small size of the air pocket and the constrictions of the cavity itself. In other cases the bubble may pop about in a lively manner as the specimen is tipped back and forth.

Quartz crystals from various localities in Brazil, particularly smoky and amethystine varieties, have been prone to contain movable water bubbles in them—as well as an incredible number of other mineral inclusions. Occasionally the water-filled cavity will enclose loose perfect crystals of the host mineral or other minerals in suspension in the water! These suspended crystals will move about in the cavity. I have examined water-filled bubbles in large quartz crystals that enclose microscopic perfect crystals of quartz. My own collection includes a quartz-crystal fragment from Brazil that contains a very large bubble and water pocket as well as some two dozen loose, microscopic perfect crystals of chlorite that tumble about in the cavity as the specimen is rotated. In another find in Minas Gerais, Brazil, I examined a 15″ smoky-amethystine quartz crystal that contained some seventy-two movable bubbles! Large, indistinct, crude, translucent crystals of quartz have also come from Brazil that can only be described as enclosing a small puddle!

There have also been some interesting reports on the mining of bubble-filled quartz crystals that were left exposed to freezing weather and actually exploded when the water froze inside the crystal.

The famous crystallized-gypsum-variety-selenite locality of Cave of the Swords, Naica, Chihuahua, Mexico has been the source of some of the most remarkable water-bubble inclusions in the mineral world. This locality, when first encountered during mining operations in the forties, consisted of a natural cavern in which there were huge clusters of long, swordlike, single and twinned crystals of transparent to translucent selenite. This was truly one of the world's greatest finds of a crystallized mineral. Many of these lustrous bladelike crystals contained one to several water bubbles as inclusions within the transparent crystals. In some cases the water bubbles had to travel a distance of over 1′ as a large crystal was tipped back and forth. Some of the more prominent water-bubble selenite crystals suggest that surely here nature has designed its own carpenter's level.

During the late sixties a pocket of complex and very beautiful cube-on-octahedron (or vice versa), clear crystals of greenish-to-colorless fluorite was found in one of the mines of Naica. Many of these crystals contained exceptionally well-defined, tiny, movable air bubbles in water-filled, tubelike cavities

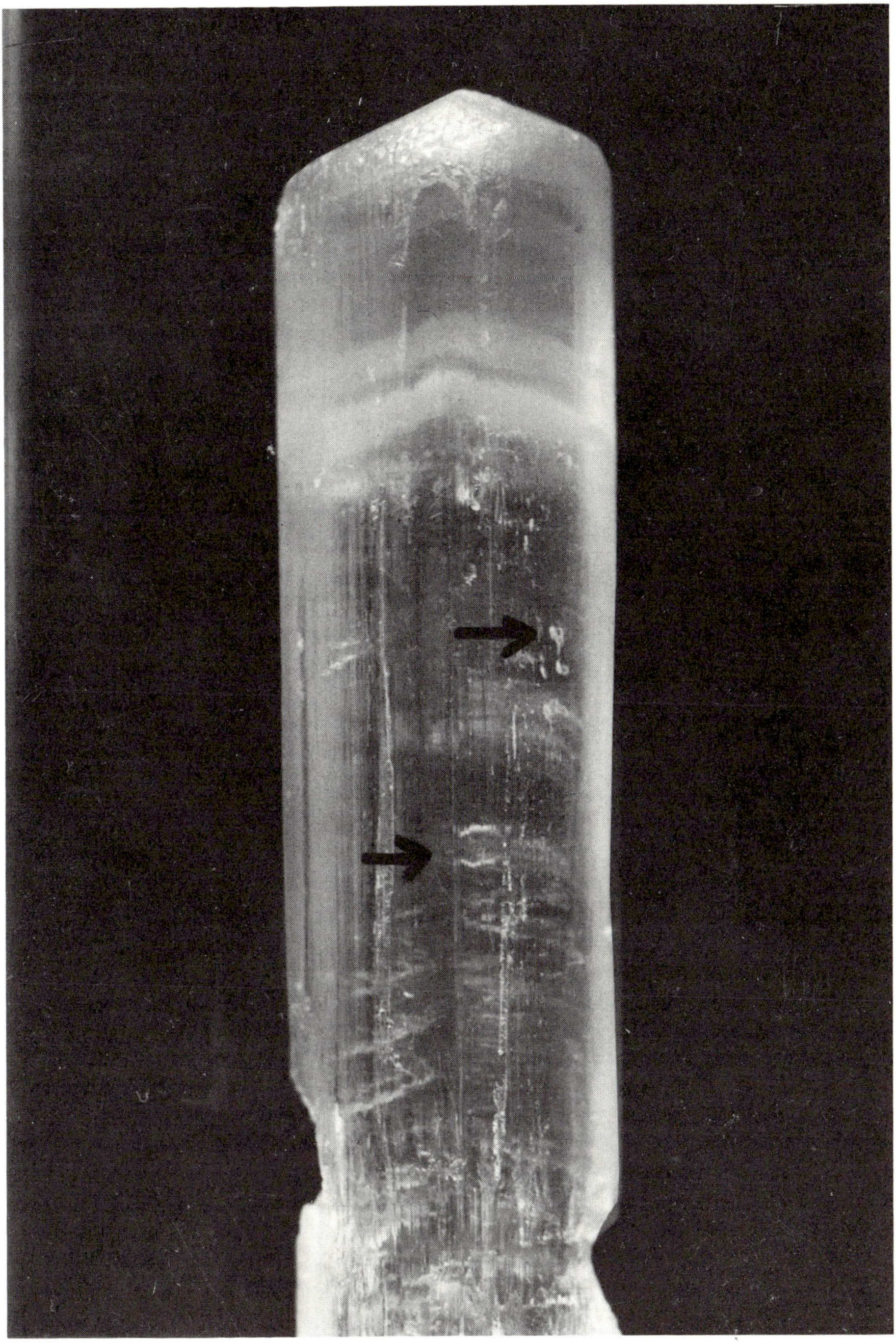

Figure 4-3. The water-bubble (air) inclusions within selenite crystals from the Cave of the Swords, Naica, Mexico are world-famed. This *zoned* crystal is 10" × 1½" and contains eight trapped bubbles that travel up and down the crystal as it is tipped. The black arrows point to two of the larger air pockets.

in the fluorite crystals. Water bubbles have been encountered in fluorite from other localities, particularly from the mines of southern Illinois and the Castle Dome District, Yuma County, Arizona, but none match the fantastic distinctness of the Naica find.

Many of the very clear, doubly terminated quartz crystals, known as Herkimer diamonds, from various localities in Herkimer County, New York provide one of the world's most incredible inclusion phenomena. Many crystals have been found at this long-known locality in which a water-bubble inclusion will also include a tiny particle of black carbon suspended freely in the water. In some specimens the carbon particle can be observed under a high-powered microscope to oscillate violently as it is bounced about by the ever-active atoms composing the water. This is known as Brownian movement, or Brownian motion, a phenomenon that has been important in science since 1827 when the noted botanist Robert Brown first discovered it.

Not all water inclusions in minerals are evident to the naked eye or even under a hand lens. The remarkable play, or flashing, of colors within precious fire opal is due to the presence of microscopic inclusions of water. Similarly, so-called milky quartz gains its opaque appearance from the presence of countless thousands of tiny water-bubble inclusions. Factors such as these, however, can only be observed with a powerful microscope.

Sand Inclusions

There are three minerals in particular—gypsum-variety selenite, barite, and calcite—which often occur well crystallized and show inclusions of varying quantities of *sandlike* material. It is not uncommon for these inclusions to dominate the specimen. The only way to identify such a specimen is by its basic crystal shape, or habit, and by other physical-identification properties such as hardness and acid tests, since heavy sand inclusions can truly mask a mineral. Specimens of this type can represent an admixture of the host crystallizing mineral and from 50 to 70 percent foreign sand materials by weight.

Inclusion specimens like these are abundant at a number of localities throughout the world. Apparently the minerals crystallized within beds of sandy or gravellike deposits. As the crystals formed, they actually involved and included the grainlike particles in the immediate area of crystallization. This might be compared to pouring liquid plastic or thin glue into a sand dune and, once the material had hardened, removing a mass now solidifying the sand granules.

Very fine examples of this phenomenon are found at several localities in the United States as well as Mexico. For example, in the topsoil surrounding the Norman, Oklahoma locality, barite roses varying from 1" to 4" in diameter have been found in great numbers for many years. They are multiple clusters of barite crystals that crystallized in a reddish sandlike formation and solidified the sand granules into the basic crystal-structure complex. These multiple groups look astonishingly like full-blown roses. Reddish-hued, perfect, sharp examples make very attractive specimens.

Gypsum also occurs frequently in roselike crystal aggregates similar to barite. Gypsum roses with heavy inclusions of sand particles have been found at numerous localities. Exceptionally attractive specimens have been collected in the Salton Sea area, Imperial County, California.

Similarly, another long-known classic locality producing spectacular specimens is the Rattlesnake Butte area of the badlands of Washington County,

South Dakota. Here calcite crystals (always doubly terminated) have crystallized in beds of exceptionally fine salt-and-pepper-like sand. Although they are completely opaque, due to the dense inclusion of the sand material, these crystal specimens show gleaming crystal surfaces under the right light exposure and can be very attractive. The Rattlesnake Butte locality and other nearby sites produce specimens measuring from 1½″ to huge, showy clusters 12″ across. Single "sand crystals," the local name for these specimens, have been found up to 6″ long. The finest specimens are found only by extensive digging, since once they are exposed on the surface or even near enough to the surface to be saturated by rainwater, the crystals lose much of their facial sharpness and begin to disintegrate. Similar sand-crystal specimens have been found near Cinco in Red Rock Canyon, Kern County, California, but they are not as sharp and distinct as the South Dakota specimens.

Another notable occurrence of sand-calcite crystals is in the limestone areas surrounding Fontainebleau and Nemours, France. Here small (up to ½″) individual rhombic-calcite crystals enclose granules of pure, white silica sand. These little crystals are completely opaque and look very much like rhombic sugar cubes.

In most of the sand-calcite-crystal localities, specimens are found in which the calcite has been almost completely leached away, leaving behind only a minute trace cementing the granules together. In such cases the specimens are actually sand pseudomorphs after calcite crystals!

Figure 4-4. Under certain conditions a mineral may crystallize within a sand environment! The crystallization process then includes the surrounding sand granules. This is a sand-calcite group from near Rattlesnake Butte, South Dakota, measuring 2½″ × 2″.

Perhaps the most fascinating of all the sandlike inclusions are the countless specimens found in excellent very sharp, bladed clusters of gypsum-variety selenite in the deposits of an ancient saline lake at a locality known as the Great Salt Plain, near the town of Jet, Alfalfa County, Oklahoma. These crystals and groups range from small 1½" specimens to large clusters 1' or more in diameter. At this locality the selenite has crystallized in the bladed habit of gypsum and included quantities of a light-brown-colored sand that occurs in the formation beds. Usually the crystals are opaque, since they are filled with sand and even small pebbles. Sometimes the sand is included as a very odd hourglass-shaped phantom in a transparent crystal.

There are at least two other similar localities in northern Mexico that have produced crystal clusters up to 2' and 3' in diameter involving single crystals as large as a man's hand. The Mexican material is usually grayish in color and opaque.

Figure 4-5. This group of sharp, very well-defined gypsum (selenite) crystals is completely opaque due to the inclusion of very heavy grains of sand. Clusters of these interesting specimens are found at various localities in Alfalfa County, Oklahoma.

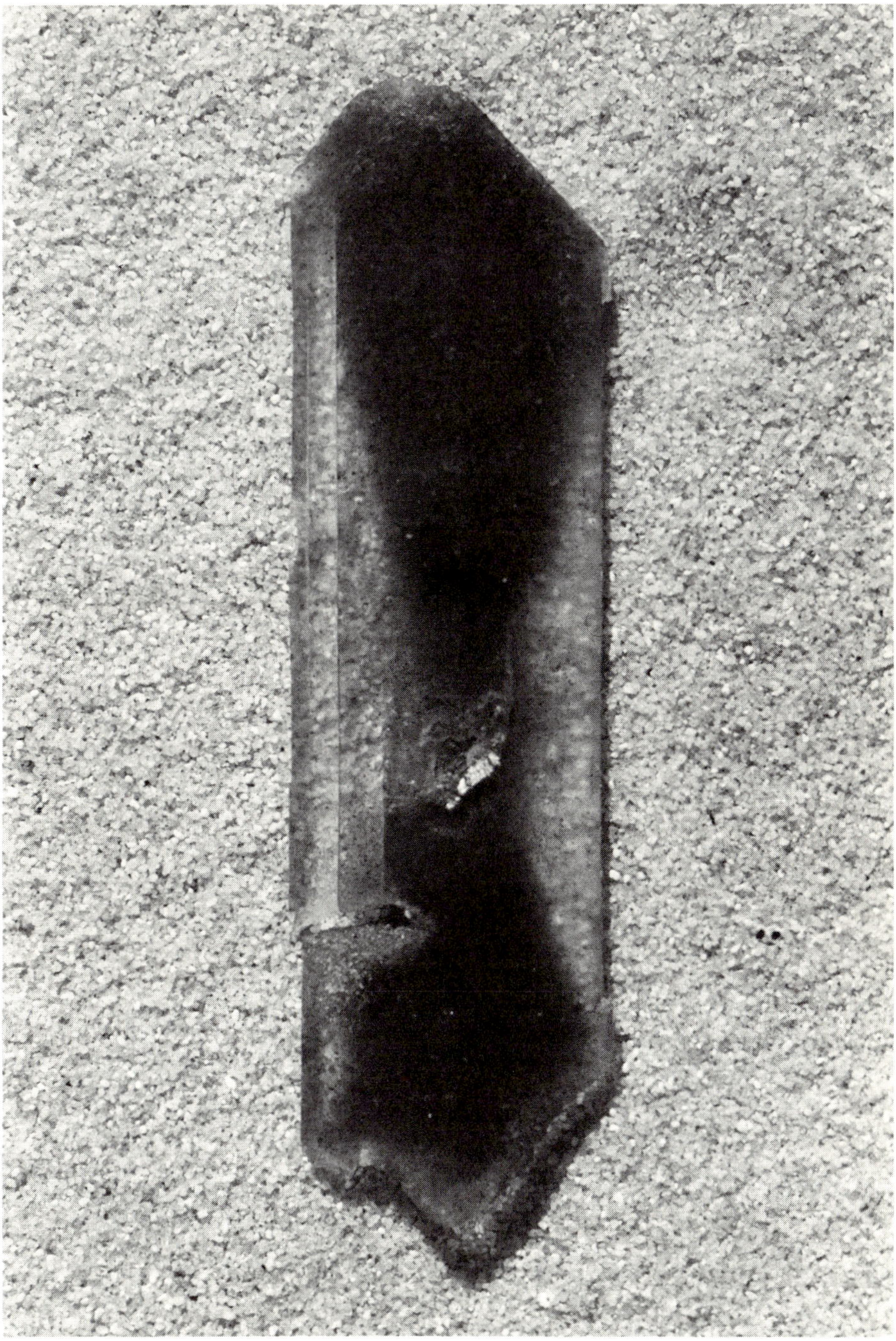

Figure 4-6. This fascinating crystal of selenite from near Jet, Oklahoma shows a phantom inclusion composed entirely of grains of sand! Selenite, or gypsum, is one of the world's most abundant crystallized minerals and is often found containing various inclusions.

Chapter 5

AGATE NODULES, THUNDEREGGS, AND GEODES

To both the novice and the advanced collector, nodules and geodes are among the most beautiful, interesting, and provocative of all mineral forms. They may be spectacular specimens.

When the term *nodule* is mentioned, most collectors associate it with agate-filled nodules, or *thundereggs*. This is quite correct. However, you should keep in mind that the term "nodule" is also one of the basic physical-description terms in mineralogy. It can be used to describe the shape of virtually any mineral occurring in round or oblate spheroids. Many minerals in nodular form have been found at various localities. It is one of the more common forms in which minerals occur.

A *geode* is simply a hollow nodule! Again, "geode" is a term in descriptive mineralogy describing a hollow nodular form or a series of crystal-lined cavities in a specimen.

Nodules and/or geodes may form in virtually any type of rock mass or ore body in which there are spherical cavities, either large or small. The most common nodules and geodes found by collectors are of silica quartz in the form of agate, chalcedony, or crystalline quartz-lined geodes.

Agate or quartz nodules are formed in several situations: in cavities within igneous-volcanic rock, created by gases as the rock cooled; in sedimentary-rock cavities, produced by gases from organic sources; or as replacements in cavities caused by disintegrated concretions. Agate can also be deposited in cracks and seams in rock as well.

What happens within the original cavity—the hollow area—to produce a geode? Mineral-laden waters percolate into the cavities and either completely or partially fill them with material that is basically *harder* than the surrounding rock. Then, as the rock formations gradually reach the surface of the earth and are exposed to the elements of erosion, the former bubbles or cavities (now very hard) weather out to rest on the ground as nodules or geodes. In all cases a hole, a cavity, in solid rock is slowly filled with mineral material. The cavities can range from the size of a pea to monstrous openings 10′ × 30′, lined with crystals.

Nodules and geodes of very large size (some up to 2′ to 3′) have been found along the Deschutes River and elsewhere in Oregon, as well as in the agate beds of the Chocolate and Chuckawalla Mountains of southern California.

Geodes, hollow nodules, are found within basic nodule-bearing beds or strata. In a sense a geode is a cavity that is arrested during the filling process.

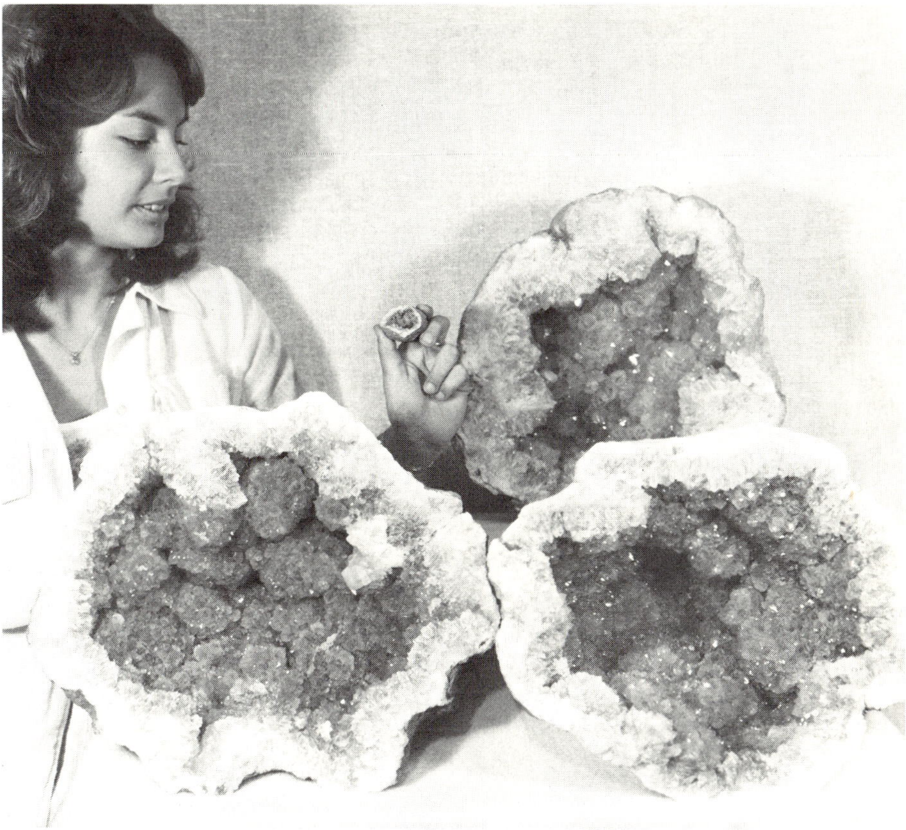

Figure 5-1. These large, very fine *geodes* are from near St. Francisville, Missouri. They are lined with beautiful calcite and quartz crystals. When they are dug out of the matrix and broken open, they are always filled with water. This locality and others in nearby Iowa and Illinois have produced some of the world's most spectacular museum-quality geode specimens.

In theory, geodes would eventually grow to become solid masses—nodules. Fine geodes make very attractive specimens. They are usually lined with quartz crystals, which may be colorless, smoky, or various shades of amethyst. Many other minerals also occur well crystallized in geodes or in geodal forms, such as calcite, dolomite, barite, selenite, celestite, and even azurite.

At most localities both geodes and nodules must be dug out of the surface of the ground, the sides of creeks and ravines, or quarry walls. In most geode-bearing beds the great proportion of nodules are solid, and geodes are found only infrequently. The proportion differs considerably from locality to locality, but very seldom is a deposit found in which most of the nodules uncovered will prove to be geodes. Why the unique growth pattern of solid nodules intermingled with a few geodes at the identical spot exists is not fully understood. Geode-bearing beds may also produce many hollow, unattractive nodules. Perhaps they are lined with a druse of dull crystals or simply with botryoidal chalcedony. The variations found within a given geode bed are often surprising.

Perhaps the most spectacular and varied geode locality in the United States is the area within a 40-mile radius of Keokuk, Iowa, embracing a number of sites on either side of the Mississippi River in Iowa, Illinois, and Missouri. The most highly prized specimens are very thin-walled and lined with sparkling crystals.

This area comprises a portion of a vast ancient inland sea. Technically, the geodes are associated with Mississippian strata, perhaps 250 million years old, and involve various limestones and dolomitic and calcareous shales. The geodes from this area occur primarily in the Lower Warsaw formation, sandwiched between the Keokuk and Salem formations, and are found in the shales as well as in the contact areas, or zones, between the limestones and shales.

Several theories have been advanced regarding the origin of the geodes of this area, and they have varied from the sublime to the ridiculous. At present the most logical theory is that the geodes themselves are replacements—perhaps pseudomorphs—of ancient concretions formed in a fairly deep water environment. The geology definitely suggests that fragments of biological material—bits of bone, shell, vegetation, etc.—could have sunk to the bottom of a sea area and attracted calcite to form a "biolith" around decaying material. These spherical concretions may eventually have been incorporated into the rock layers formed by sedimentation. By processes more or less postulated rather than proven, the concretions may have been dissolved by carbonic-acid-laden waters, creating a cavity in which silica- and calcite-laden solutions deposited various crystallized minerals in an effort to refill the opening.

Today bluffs, limestone-quarry excavations, stream-cut banks, and low cliffs expose "beds" of geodiferous-bearing shales. The nodules and/or geodes may range from the size of a bean to well over 2′ in diameter. Virtually all geodes dug out of unweathered rock strata are filled with fluid. Fluid-filled geodes reaching the surface through normal erosion processes will freeze in winter and crack or shatter. Nodules of various sizes, solid or hollow, are encountered in the same beds.

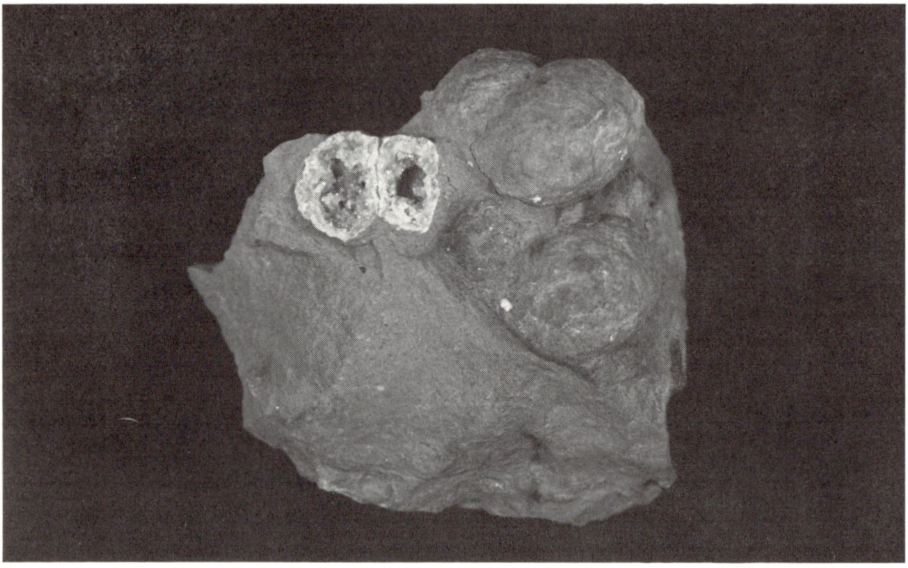

Figure 5-2. A fine-grained shale is the typical host rock in which most of the geodes are found in the Missouri-Iowa-Illinois geode region. This is a typical shale specimen showing two small, broken, quartz-crystal-lined geodes in matrix and other nodules, or geodes, to the right of them.

The Keokuk area has produced thousands of geodes for many years, although many of the most accessible localities have been exhausted or closed to the public. A large number of astonishingly beautiful geodes have come from this area, particularly from St. Francisville, Missouri. This locality is on private property and is actually a specimen mine, having produced many superb museum-quality geode specimens up to 2½′ in diameter.

Geodes from this general region, although primarily lined with quartz crystals or dull grayish chalcedony, may also contain one or more other well-crystallized minerals within the cavity. The more common minerals encountered are calcite, dolomite, aragonite, selenite, barite, sphalerite, pyrite, chalcopyrite, goethite, marcasite, millerite, kaolinite, reddish stains of hematite, pyrolusite, and, rarely, traces of smithsonite and malachite. This is quite an impressive listing, and both mineralogically and physically the best specimens from this area are very handsome and make fine collection items.

It is interesting to note that there are several markedly different localities in the general area mentioned. Some localities are virtually adjacent to each other, while others are miles away. The point involved is that *all* geode deposits display their own distinctive combination or appearance of the minerals involved. The localities of Hamilton Quarry and Crystal Glen in Illinois, for example, are only about 2½ miles apart, yet the geodes are markedly different at each site. Both localities are on private property.

Enhydros

The term *enhydros* refers to a water-filled geode. Since geodes are essentially produced by mineral-laden waters percolating into a cavity, it is not uncommon to find the growth process still taking place.

Most geode localities or beds produce dry specimens, but there are several localities in which enhydros are very common. For many years Uruguay has been the source of small, extremely thin-walled chalcedony nodules filled with water. The thin, shell-like geodes are quite translucent, and the water can be seen rocking back and forth as the specimen is tipped. The largest of these will even emit an audible splashing sound when shaken. After they are removed from the original site, these specimens will dry out in a period of several years.

The various agate-nodule beds in Brazil have also been the source of many very fine enhydros specimens. Here too specimens may be found that splash when shaken. The Brazilian specimens are usually cut very carefully on one side, then polished to produce a windowlike area showing the water-filled cavity in the agate nodule. The water can be seen swishing back and forth as the agate is tipped. Some of these may be 1′ or more in diameter. Although they make very interesting and beautiful specimens, they too are subject to drying out as the years go by.

Agates washing up on the beaches of northern California and Oregon have been found enclosing a small, movable droplet of water. They are not common and make interesting specimens when cut and polished. What you see, of course, is an air bubble in a water-filled cavity.

Perhaps the most prolific enhydros sources in the United States are the various geode beds already mentioned in the states of Illinois, Missouri, and Iowa. Several of the localities in this area yield geodes that spill water when cracked open. The most notable is the St. Francisville, Missouri bed, where very large geodes were found containing over a quart of water.

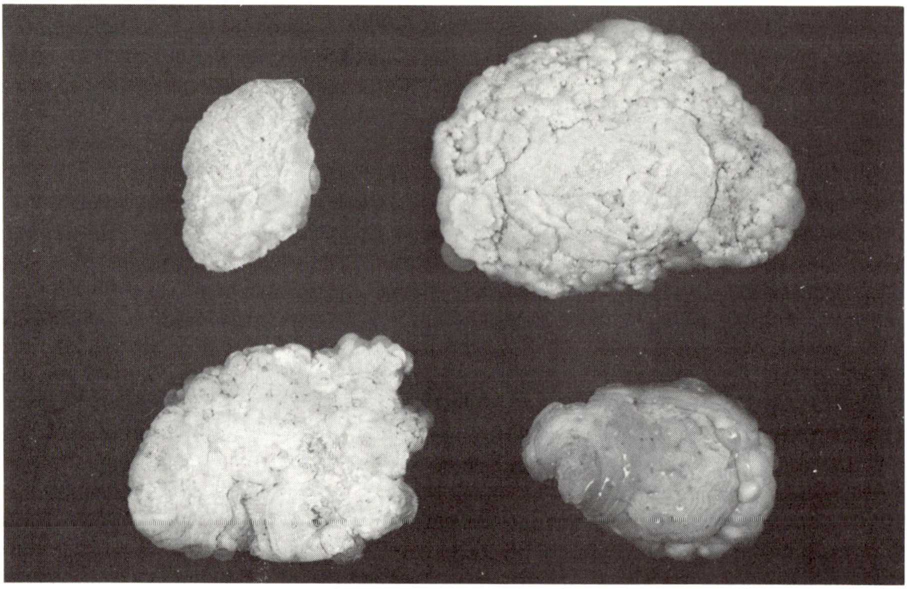

Figure 5-3. These are eggshell *enhydros*. They are very thin-walled chalcedony geodes that are partially filled with water. When they are tipped or held up to the light, the water may be seen moving back and forth. These specimens are from Uruguay and, like most enhydros, will dry out within a few years after removal from the surrounding rock.

Bitumen-filled Geodes

Now we come to the frosting—or would you believe oil—on the geode cake! I have encountered very knowledgeable mineral collectors and lapidary enthusiasts as well as professional geologists and paleontologists who have regarded the following data as unreal or some sort of practical joke.

It is a fact that there is a bed of small- to medium-sized quartz-crystal-lined geodes not more than 50′ from a main highway in Illinois that almost has to be seen to be believed! In this totally inconspicuous bed many of the geodes are filled with either a solid or liquid form of *bitumen* (or if you prefer, tar or petroleum). These are the truly odd and unique *oil geodes* found on a tributary of Tyson Creek about 1¼ miles south of Niota, Hancock County, Illinois, on State Highway 96. They are real mind-bogglers. Many of the geodes are filled with a hard asphaltlike bitumen, and others with a very thick, viscous liquid bitumen. The latter types do not splash when shaken, due to the viscosity of the liquid; however, they definitely constitute the messiest collecting imaginable! To the novice it seems incredible to dig out a hard, grayish, round rock, ring it with gentle taps of a rock hammer, and have it pop open, spilling dark black oil on your hands!

Through the courtesy of Dr. S. R. Silverman of Standard Oil of California's California Research Corporation, a very thorough analysis was conducted by their laboratories on bitumen-filled geodes I collected at this site—one with solid-, the other with fluid-type bitumen. A series of complicated chemical and isotopic analytical procedures were applied to the specimens, resulting in a lengthy and highly technical report. Some of the more interesting factors are condensed here. First of all, the geodes themselves are of inorganic origin, while their contents (the bitumen) are of organic origin. The organic substances in both geodes were completely soluble in carbon disulfide, and an infrared spectral analysis indicated that the solid and fluid bitumens were virtually identical chemically. The analysis also revealed the bitumen-paraffinic composition to be quite typical of Illinois crude oils. It was also determined that carbonate crystals (calcite) from the internal surfaces of the oil-filled geodes were of inorganic origin.

How did the oil get into the geodes? There are two schools of thought, and the question is yet to be resolved. Some geologists believe that the oil represents a small-scale local oil generation from the immediate surrounding sediments. Others support the premise that the oil emigrated from nearby oil pools or deposits. In the case of the Niota geodes, the nearest major oil accumulation, known as the Colmar-Plymouth Pool, is located some 25 miles distant at the boundary of Hancock and McDonough Counties.

Figure 5-4. Dr. E. G. Tribbey holding freshly broken petroliferous geodes at the beds on a tributary of Tyson Creek, near Niota, Illinois. This is a mind-boggling locality; where else can you break open a round "rock" and find it hollow, crystal-lined, and completely filled with black oil!

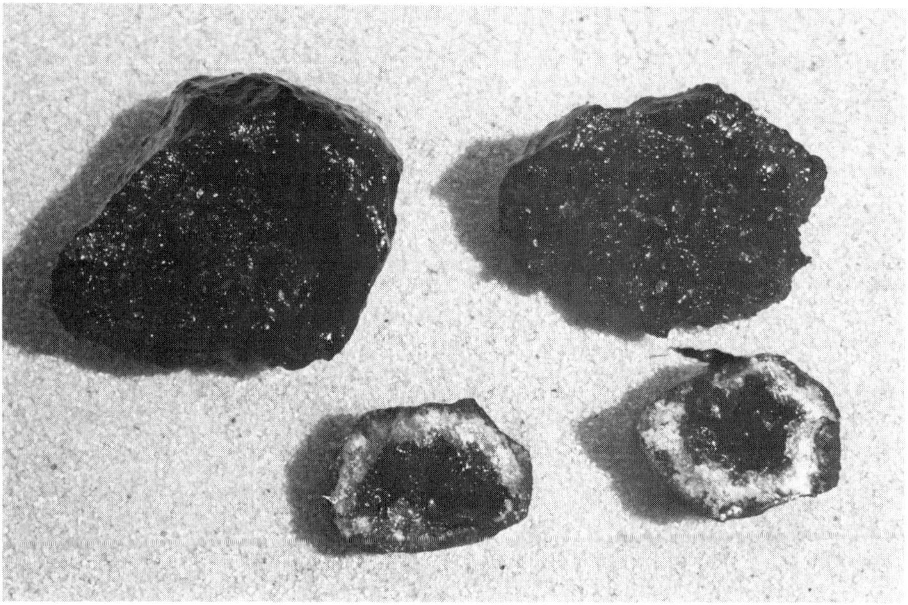

Figure 5-5. What is without a doubt one of the world's most unusual mineralogical inclusions also involves the title of the world's dirtiest collecting! The Tyson Creek geodes are basically like others in the region—thin-walled and lined with quartz, calcite, and dolomite crystals. The inclusion of viscous liquid bitumen is a mystery.

Incidentally, oil-filled geodes are not limited to this one locality: the same general region has turned up isolated occurrences, and it is not uncommon to find oil filling small solution cavities in some limestone quarries. An interesting related sidelight is that in certain pockets of fine fluorite crystals from the mines of Rosiclare and nearby Cave-in-Rock, Illinois well-crystallized fluorite often occurs completely coated with bitumen! In such cases the fluorite specimens must be cleaned with benzine or gasoline to remove the unattractive flat black coating. I have also collected calcite crystals in the Hardin County area that definitely contain bitumen flecks as inclusions.

Oil-filled geodes have been reported from the Mideast as well as Utah, but I have never had an opportunity to inspect these specimens.

Rattle Stones

In all the geode localities mentioned plus several others, geodes are sometimes found that show a stage of arrested growth. In such instances the external supply of mineral-laden waters has been terminated. Usually the solutions remaining inside the nodule deposit their materials on the walls of the nodule-geode formation, but not always! Sometimes the remaining materials form either a mudlike ball inside the geode or crystallize as loose, unattached single crystals or aggregates of crystals. In such cases the geode specimen may, upon being shaken, actually rattle! Terms such as *rattle stones* or *tinkle stones* have been applied to these geodes.

A long-known but now totally exhausted locality for walnut-size geodes in the Cady Mountains, San Bernardino County, California produced hundreds of small chalcedony geodes, *all* of which rattled markedly when shaken.

A similar occurrence was found in Maricopa County, Arizona.

It should also be noted that digging out geodes in some formations unquestionably loosens some crystals already attached to the geode walls and produces the same phenomenon. Nevertheless, the novice usually exhibits an amazed expression when ordinary-looking rock rattles when shaken!

Agates and Thundereggs

Beautiful banded, moss-filled, or plumose agates have fascinated man since the earliest days of the lapidary arts in ancient Egypt and India. And since the Middle Ages the towns of Idar and Oberstein in Germany have been identified with the finest agate lapidaries. During the last century thousands of boys grew up cherishing their prized "aggies" (agate marbles). Today the lapidary field enjoys an amazing abundance of exceptionally beautiful material easily fashioned by the amateur into cabochon-style gemstones, carvings, or cut-and-polished pieces for display purposes. Some type of agate or chalcedony material can be found in all of the continental states.

The term *agate* is somewhat confusing since it has been used in various ways and not always accurately. The terms "agate" and "chalcedony" are often used interchangeably: agate and chalcedony are essentially identical and are simply amorphous, fine-grained forms of quartz: i.e., silica.

Worldwide, chalcedony is a very common mineral, has a waxy luster, and is usually rather colorless to bluish gray. It often occurs in botryoidal linings of cavities and geodes as well as in nodules, cracks, and seams in rock. On a scale of 10 it has a hardness of 7. When chalcedony is found in translucent, colorful, banded or inclusion-filled specimens it is usually called agate. In the early days of the lapidist, fine, banded black-and-white and red-and-white agate was called onyx and sardonyx, respectively. Jasper is another form of opaque amorphous quartz, and both fire opal and common opal are members of the silica team. Common opal is a frequent filling in some of the Oregon thunderegg beds.

The frequent occurrence of fossils that are agatized, jasperized, or opalized indicates the particular type of silica replacement involved. Quartz and its crystallized and amorphous forms require well over one hundred different names for legitimately recognized species, not counting those that are purely local in nature. Several are tawdry and deceptive—quartz nomenclature is mineralogy's proverbial can of worms.

For years gorgeous banded-agate nodules and geodes have come from various localities in Brazil and Uruguay, and elliptical nodules up to 15" in diameter are not uncommon. The South American localities continue to furnish an abundance of fine material. Various localities in Texas, Idaho, and Arizona have also produced very distinctive quality material. Particularly stunning agate nodules and geodes have come from several localities in Mexico, and unquestionably select specimens rank among the most beautiful of the world's agates. Some of the Mexican nodules, particularly those from Chihuahua, are remarkably spherical and have been given the nickname "coconuts."

Of all the American localities, however, none are more provocative than the *thunderegg* beds of Oregon, Washington, California, Utah, and Nevada. The name supposedly came from the Indians around the Warm Springs Reservation in Oregon, who found beautiful agate nodules eroded out of the hills after a heavy thunderstorm and subsequently coined the term thundereggs. True or not, the term is picturesque, and there are hundreds of collectors who specialize in the cutting and polishing of "eggs."

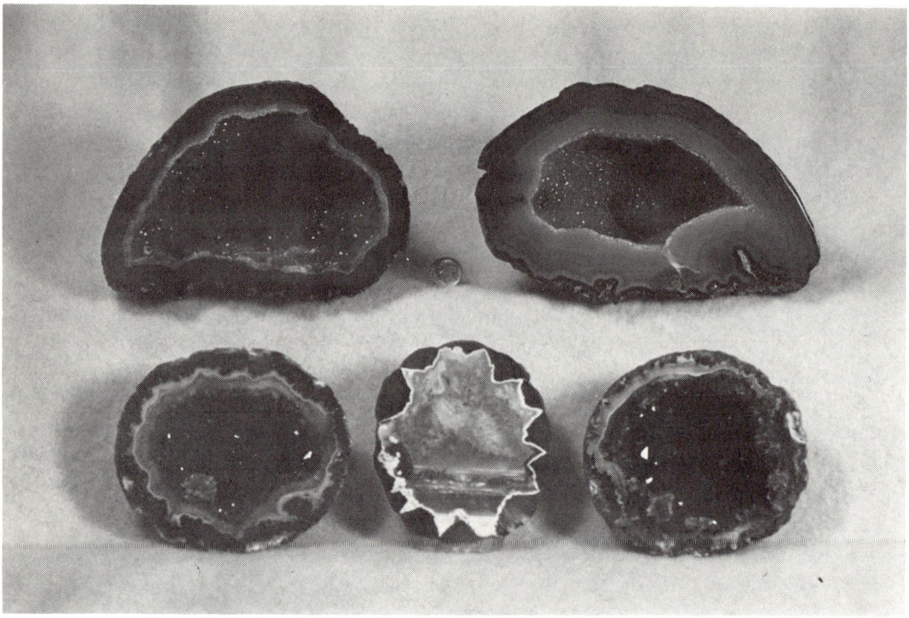

Figure 5-6. The top two specimens are fine-quality *banded-agate* geodes from Uruguay and Brazil. The left- and right-hand specimens in the lower row are from Chihuahua, Mexico and are lined with gem-quality amethyst. The *thunderegg* in the center is typical of many nodule beds in Oregon. This specimen is from near Burns.

The deposits of agate- or chalcedony-filled nodules are usually located in an easily defined, fairly level stratum or series of adjacent strata. Consequently, the term *beds* has been used to identify a specific locality, such as the Priday Ranch Beds, the Antelope Creek Beds, or the Hauser Beds.

Many of the former thunderegg beds in the West are virtually exhausted. They have been picked clean, pitted, dug, trenched, blasted, and even completely wiped out by the commercial collector and the bulldozer. Some beds look like World War II battlefields. There are, however, commercially operated thunderegg beds where the owners use heavy equipment to remove the overburden so that collectors may have access to nodule-producing strata.

Oregon continues to be the source of hundreds of fine specimens every year from a variety of localities more or less centered around Madras in the central-eastern portion of the state. Many visitors come to Oregon specifically to collect agate nodules, and data on localities is readily available. Many of the better and more productive sites are on private property, and nominal fees are charged for digging privileges. Several of the thunderegg-bed owners advertise regularly in the mineral and lapidary publications, giving details as to the dates the beds are open for digging, the fees, and the type of material currently being found.

It should be noted here that not all experts agree on exactly how agate nodules of the thunderegg type form. That they basically develop in gas-created cavities in formerly viscous volcanic rock, such as rhyolitic lavas and tuffs, seems well established. However, some believe the agate deposition is a constant process of internal encrustation and the eventual filling of the cavity by silica-laden percolating waters. Another theory contends that the cavity was filled *completely* at some time with a gelatinous solution of silica gel and that this, plus impurities, formed the colorful interior fillings of the thundereggs. Evidence indicates that the silica is of hydrothermal origin and was deposited in the cavities while the rocks were still hot or warm. Specimens that tend to confirm *both* theories are abundant.

A fascinating and disturbingly unexplained factor of thundereggs involves the starlike, or radically pointed, configuration of almost all agate fillings within these nodules. No matter how an egg is cut, this format is evident. The nodule may be round, oval, or even double, but the agate filling itself will comprise about two-thirds of the nodule in a pointed starlike shape, with the remainder of the nodule matrix usually a brown or greenish material. At some beds this internal filling may also be more or less cubic. It has been theorized that these patterns represent rupture marks of an originally angular cavity. The expanding gases in the viscous lavas may have produced a starlike or crudely cubic cavity that later developed various degrees of symmetry. Regardless of the thousands of thundereggs that have been cut and polished, the precise explanation for this amazingly consistent configuration is not known.

Agate nodules are often found in which the center fillings are obviously composed of different types of agates—of various porosities—which must have been deposited in a sequential manner. Of particular interest are fillings in which straight horizontal lines are present, often in attractive scenic patterns. Frequently a set of these lines will be found tipped at different angles from the rest, which clearly indicates that the bed of thundereggs has been subjected to major earth movements. I have examined eggs from both the Priday Beds in Oregon and the Chocolate Mountains in California that showed perhaps four different angled layers of agate.

A somewhat related and interesting phenomenon are nodules that show positive evidence of being cracked in half and then resealed. There are two forms: one, an empty cavity falling on a rock fracture line that is later filled with agate; and two, a developed agate nodule that has been cracked by earth movements and has resealed itself. In both cases there is an observable "cracked" offset to the specimen.

Certainly among the most attractive specimens of the mineral world are the agates with beautiful inclusions. Agate nodules from selected localities have been cut and polished to show incredibly realistic "scenes," such as cloud formations, ocean waves breaking on a beach, tree-studded islands in a quiet sea, mountain peaks soaring into puffy clouds, human and other animal shapes, brilliant-colored flower gardens, or undersea scenes.

Each agate is truly individual: no two are ever perfectly alike. The good fortune of cutting a specimen so that a scene shows is essentially a matter of luck. Of course, there is also a certain degree of skill involved. Knowledgeable lapidaries know the depositional characteristics of certain agate beds and are familiar with preferential angles to cut a given specimen in order to achieve the highest possible degree of plumes, pompoms, scenic layers, or concentric rings.

Although it is seldom realized, the fundamental material involved in the attractive banded, moss, plume, or scenic agates is the product of different foreign elements being *included* during the deposition of the agate or of chemical coloring agents penetrating the agate after formation. Chalcedony is essentially colorless, but this condition can be changed if, during deposition, various minute portions of other minerals are held by the silica solutions, thus producing various growth patterns much like the growth-deposit rings in a stalactite or stalagmite.

A fine Fairburn, Great Lakes, Chihuahua, or Brazilian banded agate is the product of different chemicals included with the basic silica at various times as the nodule is in the process of formation. Sometimes these are referred to as "impurities," but whatever you wish to call them, they are colored chemical inclusions within essentially colorless silica that have produced the beautiful multicolored, banded, and scenic agates sought by collectors and lapidary enthusiasts.

There are literally dozens of local names given to agates from different localities, and they do much to help define the unique type, or pattern, of the agate involved. Again, the distinctive growth patterns are all the product of different types of chemical inclusions characteristic of a particular locality. For example: Priday Plume, Crazy Lace, Montana Moss, Fairburn, Laguna Banded, Brazilian Banded, Graveyard Point Plume, Mint Canyon Fortification, Texas Plume, Owyhee Picture, Mexican Fire, Indian Green Moss, Pauline Red Moss, Glendive Red, Kramer Hills Dendritic, Pipe Springs Blue, and Bruneau Jasper are a few typical local identification names given to highly distinctive agates or jaspers. The degrees of distinction vary from localities producing inclusion patterns of extreme delicacy, such as some of South Dakota's fine banded Fairburn agates and the flowerlike plumes from Texas and Oregon, to heavy banded, contrasting layers, characterizing many Brazilian agates.

Figure 5-7. Agate nodules are infrequently found showing evidence that the original cavity had been fractured and later filled with chalcedony. Nodules are also found indicating that the specimen was fractured by earth movements during growth and resealed itself. This cracked-and-resealed nodule is from Mint Canyon, California.

Experts as well as amateur hobbyists can become well acquainted with different agate localities and their distinctive identification qualities. They will have no more trouble identifying a nodule from the Priday Ranch in Oregon as opposed to one of the Chocolate Mountain nodules than an experienced mineral collector has in differentiating smithsonite from the Kelly mine in New Mexico from smithsonite of Berg Aukus, South West Africa. To the sophisticated collector of agates and related amorphous quartz-family minerals, the distinctive localities are as important as the specimens themselves!

The foregoing information refers to natural specimens (which can be of every color of the rainbow), *not* to artificially dyed agate specimens. The technique of dyeing agates by various chemical and heating techniques was known to the Romans and by the early 1800s had been developed commercially in Germany. In such cases man introduces an inclusion into the original, often unattractive material involved. A structurally banded but essentially colorless agate is soaked in various solutions, either hot or cold, sometimes pressurized, and often heated in sand-filled retorts under controlled conditions.

Agate is really not solid at all! The depositional growth layers of the agate reflect different degrees of porosity and will accept chemical coloring agents in varying degrees. The artificial coloring or dyeing of agates is done after they have been cut and polished. A typical procedure involves soaking a polished stone or slab in a saturated solution of warm sugar water for several weeks. The sugar penetrates different layers of the agate in various degrees, then the specimen is washed, placed in concentrated sulfuric acid, and heated. This carbonizes the sugar and turns the agate black or brown, depending on the porosity of the layers affected.

Coating sections of the agate with wax and then soaking the stone in various chemicals can also produce a wide range of colorful artificial patterns. Usually the coloring is only superficial, penetrating the agate for 1/100″ or so, again depending on the porosity of the particular agate involved. The colors may be tasteful or garish.

Chapter 6

PSEUDOMORPHS—THE GHOSTS OF MINERALS PAST

If Charles Dickens had been a mineralogist, he might have referred to pseudomorphs as "the ghosts of minerals past." If you combine the Greek word *pseudo*, which means "false," with *morph*, which means "form," the product is a *pseudomorph*, or false form. These specimens are among the most intriguing and fascinating for both the amateur and the sophisticated collector.

In essence, a pseudomorph represents the replacement or alteration of one mineral by another while retaining the basic shape or form of the original. This relationship is usually referred to in mineralogical jargon as "one mineral *after* another," for example, goethite after pyrite, chalcedony after barite, hematite after magnetite, malachite after azurite, or gypsum after glauberite. There are literally dozens of examples.

Figure 6-1. These specimens are *pseudomorphs:* native copper has replaced all of the former material. At the top is a fossil tree limb from Bisbee, Arizona; below it to the left and center are two azurite rosettes from Santa Rita, New Mexico; at the lower right is a group of aragonite crystals from Corococo, Bolivia. Although these specimens look unattractive, they are very choice collector's items.

Most mineral collections contain various examples of pseudomorphism. Some examples are fairly common and easily identified, but others can be very obscure and discernible only to the sophisticated mineralogist. Many common pseudomorphs may go unrecognized by the beginning collector, while many advanced collectors may specialize in them! Pseudomorphs are doubly fascinating as mineral specimens since they always tell a twofold, or sometimes threefold, mineralogical story.

Pseudomorphs can be of more than casual importance or curiosity to mining engineers, mine operators—and, of course, stockholders as well. The mineralogical history of an ore deposit, or ore body, is referred to as its *paragenesis.* Essentially this means the scientific determination of how the deposit was created and in what sequence the mineralized components were deposited. All ore bodies have different origin stories, or, to put it another way, there would be little similarity between the paragenesis of the commercially important copper deposits in Tsumeb, South West Africa; Bisbee, Arizona; or Santa Rita, New Mexico. In one sense, paragenesis might be thought of as the diary of an ore deposit and consequently, it can be of enormous importance in estimating its extent and value. For this reason the presence of pseudomorphs is carefully studied, since they can provide key clues regarding which minerals were deposited first, second, and third and which major changes may have affected the ore body. Often the professional mineralogist ends up playing detective.

Are pseudomorphs abundant? This is a relative question. As is discussed in detail later, most pseudomorphs are solid physical replacements, and, certainly in some areas, fossils (many of which are true pseudomorphs) are very abundant! Insofar as crystallized minerals are concerned, most major ore deposits contain materials showing varying degrees of pseudomorphism. Remember: the phenomenon can be difficult to identify. There is no rule of thumb.

Some fine pseudomorphs from well-known localities are very scarce, yet in other localities the entire deposit may be composed of pseudomorphs! For example, the well-known Pelican Point locality in Utah, which has furnished thousands of limonite (or goethite) pseudomorphs after pyrite, is comprised of virtually nothing but pseudomorphs. Much the same situation exists at the gypsum-after-glauberite locality near Camp Verde, Arizona, and I have been on mine dumps in the Hesperus Basin of southern Colorado where most "crystals" picked up on the dumps were, in fact, some type of pseudomorph.

It would be difficult to find any museum or advanced private collection that did not display some specimen upon which the label indicated "something *after* something." Although pseudomorphs can occur in any type of mineralized area, there are a few more or less "classic" specimen localities, some of which are described as follows.

Perhaps the most forthright and appealing of well-known examples are the superb malachite-after-cuprite specimens from Chessy, France. These remarkable single-crystal pseudomorphs vary from ¼" up to 1½" in size and are very attractive specimens. In these the bright green malachite has totally replaced cubic, octahedral, and dodecahedral crystals of ruby-red cuprite and at the same time has faithfully identified cuprite crystals of exceptional size and perfection. These are choice collector's items and are now somewhat difficult to acquire.

During the last few years remarkable chalcedony pseudomorphs after very sharp and well-defined sixlings of aragonite have come from Chihuahua, Mexico. The mines at Corococo in Bolivia, South America have produced exceptional pseudomorphs of native copper after small clusters of aragonite crystals; similarly, Georgetown, New Mexico has yielded some fine native-copper pseudomorphs after azurite rosettes. Perhaps the most interesting of all copper pseudomorphs came from the Copper Queen mine in Arizona, in which a few unique small fossil-tree limbs completely replaced by native copper were found.

For many years countless gypsum as well as calcite pseudomorphs after glauberite have come from an ancient Indian salt-mine locality very close to Camp Verde, Arizona. The specimens have varied from very sharp ½″ crystals up to divergent clusters of 4″ to 6″.

With no exaggeration, there are dozens of localities that have produced numerous specimens of limonite (goethite) after pyrite, perhaps the most common ore-body pseudomorphic form that the collector may find. Pelican Point, Utah; Baden Lake, North Carolina; Pretoria, Transvaal; Sulzer, Alaska; Chihuahua, Mexico; Quartzsite, Arizona; Salzburg, Austria; and the Island of Elba, Italy are only a few of the many localities that have produced specimens of note. Some limonite-after-pyrite pseudomorph specimens may show a tendency to crack.

Both Naica, Chihuahua and Ojuela, Durango, Mexico have produced some handsome, varied limonite (or goethite) pseudomorphs after selenite. Usually these are in the form of dark brown encrustations featuring a hollow cavity in the exact shape of the vanished selenite crystal. Mexico has also yielded a magnificent pseudomorph in the form of the stibiconite replacements after stibnite from Catorce, San Luis Potosi. These are truly remarkable specimens that preserve in the utmost detail all aspects of the former stibnite crystal, including complicated striations.

There are several localities in Colorado where quartz has replaced barite, as either a hollow encrustation over a vanished barite crystal, a complete solid replacement of the exact crystal per se, or a solid replacement with *added* quartz crystals around the entire periphery of the replaced barite crystal. Some of these specimens are large and exceptionally showy.

There are numerous other localities that have produced a wide variety of pseudomorphs: various Canadian and New England sites wherein garnets have been replaced by chlorite; many localities where various feldspars have been replaced by the claylike mineral kaolinite; pegmatite areas where tourmaline, beryl, and spodumene have been replaced by mica; and localities where either limonite or goethite have replaced pyrite, barite, chalcopyrite, siderite, fluorite, magnetite, selenite, dolomite, hematite, or calcite.

The famed traprock areas of New Jersey, New York, and Connecticut have furnished some of the world's truly beautiful pseudomorphs. The mineralogy of these formations is very complicated; however, they have produced many different types of pseudomorphs since their discovery. Prehnite-encrustation pseudomorphs after what appears to be anhydrite are particularly choice, and many prehnite specimens from this area show this. Just look for the square or rectangular cavities on the reverse side of the specimen. This series of localities has also produced some magnificent "skeletal"-encrustation pseudomorphs of various minerals—quartz, prehnite, calcite—after anhydrite and other minerals. Many of these are exceptionally fragile and beautiful specimens, admired not only for their intrinsic beauty and appeal but also for the multiple pseudomorphic story they tell.

A superb example of very complex mineralogical pseudomorphism exists in the stunning azurite, malachite, and selenite combination specimens from the

Figure 6-2. This very showy specimen that looks like a large sugar cookie is a quartz pseudomorph after barite from Howardsville, Colorado. Although the quartz replacement and encrustation is quite thick, the center of this 2½" × 1¾" specimen is hollow—a reminder of the former barite.

old Apex mine near St. George, Utah. Here pseudomorphism has run wild, with colorful and delicate pseudomorphic encrustations of blue azurite and green malachite replacing themselves as ghostly encrustations after fingerlike crystals of selenite—or fantastic blue and green pseudomorphic crystals penetrating clear masses of selenite. Some of the Apex-mine pseudomorphs are positively spectacular, although the mine produced relatively few specimens that ever reached collectors or museums.

There are basically four types of pseudomorphism: encrustation; alteration; replacement, or substitution; and paramorphism. Their essential differences are briefly described as follows.

Encrustation Pseudomorphs

Encrustations are usually the most readily identifiable pseudomorphs. They are caused by one mineral coating, or *encrusting*, another mineral that is already crystallized. Minerals coating one another are assuredly not a rarity; however, in this case the first mineral (the one being encrusted) disappears. This action leaves behind a hollow shell showing the crystal shape of the original mineral but now actually composed of another.

The hollow shell of a true encrustation pseudomorph is actually a negative cast of the original mineral and is sometimes called a mold, but this is not the best of terms.

Encrustation pseudomorphs can be paper-thin or thick and massive, depending entirely on the conditions in a particular deposit. I have collected large plates of extremely delicate, tissue-paper-thin encrustation pseudomorphs of pinkish quartz after barite in Colorado's La Plata Mountains, while similar material from near Howardsville is very heavily encrusted and thick, often resembling large rectangular sugar cookies.

Figure 6-3. These are limonite, or goethite, *encrustation* pseudomorphs after a group of selenite crystals from Naica, Mexico. This black and brown specimen is composed of a series of hollow forms that clearly show the crystal shape of selenite.

Magnificent, showy, shell-like encrustation pseudomorphs have long been found in the various traprock-quarry localities of New Jersey. Of particular pseudomorphic interest have been the wafer-thin quartz-encrustation pseudomorphs after large anhydrite crystals. These same localities have provided many handsome pseudomorphs of this type, with prehnite featured as the encrusting agent. Both the quartz and the prehnite pseudomorphs from these localities are often associated with calcite and other zeolite minerals such as stilbite, apophylite, and chabazite.

Many museums as well as private collections display a true classic of pseudomorphism that was encountered many years ago in the Handsome Nea mine, Nenthead, England. These specimens are encrustation pseudomorphs composed of a *series* of interstacked thin cubic shells of pyrite crystals that had originally coated fluorite cubes. The fluorite is gone, but its ghost is well preserved by the nests of pyrite shells.

Skulls

A unique and fascinating type of pseudomorph originating as an encrustation are the native-copper *skulls* found many years ago in the various associated iron-copper localities of the Keweenaw Peninsula in Michigan.

These odd pseudomorphs are the result of arborescent sheets of native copper forming around individual rock pebbles in an ancient glacial conglomerate. After encrusting the pebble with a protective metal jacket, the host pebble disintegrated, leaving behind a ghostly, hollow metal "skull."

These geodelike pseudomorphs range from about the size of a half-dollar to as large as a basketball and may weigh a few ounces up to many pounds. Some of the very thin-walled native-copper skulls, especially those from near Houghton, will ring like a cowbell when tapped with a pencil.

Figure 6-4. Brilliant pyrite crystals compose this fine encrustation pseudomorph of pyrite after fluorite. The original fluorite has disappeared, leaving behind its cubic shape in ghostly shells of pyrite. There are also secondary crystallizations of pyrite and siderite on the specimen. It was found around the turn of the century in the Handsome Nea mine, Nenthead, England and measures 2¼" × 1½".

Figure 6-5. Ranging in size from a golf ball to a wastebasket, copper *skulls* are among the most unusual of all pseudomorphs. They are compact arborescent sheets of native copper that formed *around* pebbles in ancient glacial deposits. Eventually the pebbles disintegrated, leaving behind a "skull" of encrusted native copper. This specimen measures 4½" × 3" and came from near Houghton, Michigan.

Alteration Pseudomorphs

A very common type of pseudomorph is the result of chemical *alteration* within a related mineral family. This may involve the partial or complete replacement of one mineral by and/or with another. In this type of pseudomorph the original crystal form remains, but chemical action has replaced or substituted the original mineral with another.

Alteration pseudomorphs are commonly found within related families of minerals such as the zeolites or within closely related chemical groups such as lead, iron, or copper minerals. A very common example, perhaps the most common of all pseudomorphs (other than fossils), is the replacement of pyrite by the related iron-family minerals limonite or goethite. There are dozens of localities in the world that have produced perfect, sharp pseudomorphs that are the result of this particular combination. Incidentally, pseudomorphism is not confined to small specimens. I have examined pyrite-goethite crystal pseudomorphs—occurring in magnificent specimens. Both malachite and showing faces up to 4″ on a side!

Other common examples of more or less family-related alteration pseudomorphs are: anhydrite after gypsum, gypsum after glauberite, kaolinite after orthoclase, chlorite after garnet, and specimens of malachite after either cuprite or azurite. These are perhaps the most frequently encountered pseudomorphs—occurring in magnificent specimens. Both malachite and azurite are basic carbonates of copper that are found at numerous worldwide localities. Malachite frequently occurs in massive and botryoidal forms; however, malachite crystals are quite rare and are usually short, prismatic or needlelike emerald-green tufts. Azurite, on the other hand, is found in a wide variety of

Figure 6-6. Among the most common and abundant pseudomorph specimens are limonite, or goethite, *alteration* pseudomorphs after pyrite. In such cases a brown, hydrous iron oxide replaces the brassy pyrite. Dozens of localities have yielded good specimens. These examples, from top to bottom, left to right, are from Utah, California, Georgia, Alaska, Italy, and Austria.

habits, often highly modified. Beautiful deep blue azurite crystals may vary from microsize to single crystals as large as a hand. Pseudomorph specimens of deep green malachite are found in very well-defined, small to large crystals. This is impossible for malachite crystals themselves, but not for malachite as a pseudomorphic replacement of azurite. Fine examples rank among the most beautiful and interesting of all mineral asbestos.

The azurite-malachite relationship is common in pseudomorphs and often is not complete: i.e., specimens may show partial replacement of the azurite by the malachite. Specimens of this type are exceptionally attractive, displaying a brilliant combination of bright green and deep azure blue.

Replacement or Substitution Pseudomorphs

Analogous to alteration pseudomorphs are those in which the replacing mineral has no chemical relationship to the original mineral but simply fills in a cavity or replaces a form, yielding a perfect positive cast of the former mineral form. In the case of quartz, some well-known pseudomorphic localities involve specimens of quartz after calcite from France, after anhydrite from New Jersey, after dolomite from Montana, and after fluorite from Cuba. The semiprecious gemstone mineral crocidolite, or tigereye, from Griqualand, South West Africa is actually a remarkable pseudomorph, a quartz replacement after asbestos. In this instance the mineral quartz is a pseudomorphic *replacement*, or *substitute*, for the original amphibole mineral asbestos.

Other somewhat dramatic, showy, and very valuable examples are the specimens referred to by Australian opal miners as "pineapple fire opal." Actually these gemstone-quality specimens are pseudomorphic fire-opal replacements of clusters of glauberite crystals.

Figure 6-7. The specimen on the left appears to be a very sharp gypsum rose. It was at one time, but it has been completely *replaced* by translucent chalcedony. These fine pseudomorphs are found near Crawford, Nebraska. On the right is a gem-quality carnelian replacement after a barite rose from Cisco, Utah.

Fossils

Although seldom thought of in this specific regard, many fossils (but *not* all) are in fact pseudomorphs: in other words, they are replacements.

A *fossil* may be regarded as any evidence indicating some form of ancient prehistoric life, ranging from a sponge to a dinosaur. There are many *types* of fossilization processes that produce a wide variety of different specimens that today we can recognize as fossils. Fossils may be imprints in mud of the footprints or tail tracks of ancient reptiles. These imprints may be negative or positive: i.e., hollow depressions or filled-in areas. Fossils may also be microthin lacy, carbonized imprints of leaves, fish, insects, flowers, or fronds encased in rock-forming material. Fossils may be in the form of internal or external casts of prehistoric life forms, or they may be an open, hollow mold—a cavity indicating that some plant or animal was entombed and vanished due to decay. They may also be examples of simple preservation, in which, due to exceptionally unusual conditions such as a dry, dusty deposit, asphalt springs, or peculiar freshwater and saltwater sands, the remains have been preserved but not changed. The sloth bones from Gypsum Cave, Nevada, the thousands of bones from the La Brea tar pits, Los Angeles, California, and the amazing complex of both seashells and sea and terrestrial mammals from the Eastbluff fossil beds, Newport Beach, California are good examples.

Figure 6-8. A remarkably accurate *permineralization*, or petrification, of fossil coral from northeastern Missouri. The former coral structure has been completely replaced in perfect detail by solid, hard, pinkish-colored quartz.

The more common interpretation of the term "fossil" denotes something completely replaced (petrified) by some mineral material. Without a doubt fossils constitute the most abundant of all pseudomorphs, but in this case instead of one mineral after another, it is a mineral replacing some sort of life form—bone, tooth, shell, coral, or wood, for example.

Perhaps the most interesting mineralogical aspect of fossils as pseudomorphs is the number of minerals that have been replacement agents. These minerals may be found as detailed cellular replacement or as negative or positive casts. The most common replacing mineral is quartz, and fossilized (petrified) wood is perhaps the most abundant of all fossil materials. Wood, bone, and shell are often found replaced by quartz-family minerals of various types, such as agate, carnelian, chalcedony, jasper, and common opal. Precious opal is found replacing wood in the Virgin Valley of Nevada, and superb gem-quality fire opal replaces shells and bone fragments in the Australian opal fields.

In addition to the frequent silicification by quartz, calcite is perhaps the next most common replacement mineral, followed by barite, dolomite, marcasite, and pyrite. In the early 1900s the famed Kelly mine in Socorro County, New Mexico was the source of a few spectacular fossilized mollusk shells and crinoid stems that had been completely replaced with apple-green smithsonite of the finest quality. Other unique pseudomorph fossils include the small tree limbs totally replaced by solid native copper from Bisbee, already mentioned, and the handsome brachiopods, snails, and other types of shells replaced by marcasite or pyrite from various localities in Ohio and Illinois.

Figure 6-9. These are ancient fossil brachiopods and gastropods from Ohio and Illinois. As fossils they are very well preserved and interesting, doubly so because they have been replaced by the brassy, metallic mineral pyrite. They are actually pyrite pseudomorphs after prehistoric life forms!

Paramorphs

Perhaps the most obscure and certainly the most difficult of pseudomorphic forms to identify are the *paramorphs*. These occur only within polymorphous related minerals—those very few minerals that are chemically identical but may structure themselves internally in different ways.

In a paramorph the basic chemical composition remains absolutely unchanged, but the orientation of the internal molecular structure changes. As a rule, only refined laboratory x-ray techniques can positively establish the existence of paramorphism. In one sense, paramorphs might be likened to identical human twins, but with each showing a markedly different personality under analysis.

A very well-known paramorph, often seen in collections, is the interesting rutile paramorph after brookite from Magnet Cove, Arkansas. Brookite and rutile are identical in chemical composition: both are titanium dioxide, TiO_2. At this locality fine crystals of lustrous, dark reddish-brown "brookite" are found, but under laboratory analysis the molecular structure is revealed to be rutile—a paramorphism. The mineral anatase (formerly called octahedrite) also shares the TiO_2 formula and has been found polymorphous (or trimorphous) with both brookite and rutile.

Another example of paramorphism is seen in specimens of aragonite that have changed to calcite. Here again, the chemical composition is identical: both are calcium carbonate, $CaCO_3$, with calcite crystallizing in the hexagonal system, and aragonite in the orthorhombic. Under certain temperature conditions the aragonite, which is considered a somewhat less stable mineral than calcite, will reorient its internal ionic structure to that of calcite.

Are pseudomorph specimens valuable? This is a question for which there is really no precise answer. Many pseudomorphic specimens are basically unattractive. In other words, they may not be very showy and may have little display appeal. Nevertheless, there are many collectors who specialize in pseudomorphs due to the interesting scientific story they tell.

Pseudomorphs add a degree of mystery and a dimension of flexibility to the study and collection of minerals. Learning to recognize them is fun! No matter how experienced the collector or museum curator, there is always some pseudomorph that poses a problem as to its origin. Pseudomorphs will keep you alert!

Chapter 7

MINERAL LUMINESCENCE

Few aspects of mineralogy are more spectacular or amazing at first sight than the phenomenon of mineral luminescence. *Luminescence* embraces both fluorescence and phosphorescence.

Are luminescent minerals common? The answer is relative—yes and no. It depends almost entirely on specific localities. There are many minerals that show varying degrees of fluorescence or phosphorescence, ranging in observable intensity from very weak to amazingly brilliant. However, of the some two thousand known minerals only a very small percentage will show some degree of fluorescence or phosphorescence.

What may appear to be an ordinary garden-variety rock—dull, uninteresting, and unattractive—may suddenly be transformed into a dazzling, astonishing display of brilliant colors, as if by magic. The secret rests with certain highly selected mineral specimens (from very special localities) and the effects of ultraviolet-light wavelengths.

For example, under normal sunlight a specimen of calcite, esperite, willemite, or franklinite from the famed localities surrounding Franklin and Sterling Hill, New Jersey looks exactly like a common chunk of granite. However, the fun begins when the specimen is taken into a darkened room and exposed to ultraviolet short-wave light. The unattractive rock suddenly becomes alive with color—vivid red, brilliant yellow, and vibrant green, interspersed with black flecks! When the light source is turned off there will often be a momentary holdover of bright orange-red and perhaps a lingering pale green phosphorescence, which gradually disappears in a few seconds. To say the least, such specimens are spectacular—as long as they are viewed under ultraviolet light.

Ultraviolet Light

What is *ultraviolet light?* It is one of the various types of radiant, or electromagnetic, energy that constantly bombards our planet. We cannot see these rays. The spectrum of radiant energy is vast, beginning with the most powerful electron and particle showers from outer space, called cosmic rays. Following these in decreasing intensity are gamma rays, x rays, ultraviolet, visible light, infrared, microwaves, radio waves, ultrasonic, and sound waves. Actually, the display of varied light that we see as our rainbow of colors encompasses the very narrow band of radiant energy to which that incredible mechanism, the human eye, is sensitive.

The colors observed in fluorescence or phosphorescence are cold, harmless, and actually generated within the specimen that is exposed and consequently activated by ultraviolet light. In other words, it is a true emitted light, not a reflection.

By using special devices, instruments, or materials, we are sometimes able to "see" or feel the expression or reaction of normally invisible energy wave frequencies, either directly or indirectly. Within the basic radiation categories mentioned there are dozens of different, highly sophisticated divisions, but in ultraviolet light there are only two types—short-wave and long-wave ultraviolet-light radiation. Both are virtually invisible.

Short-wave rays are the furthest away from our visible-light spectrum. They are shorter than sunlight rays and produce the most spectacular reaction in selected mineral specimens. Approximately ninety percent of all fluorescent minerals show best under this frequency.

Short-wave energy is also responsible for sunburn: a severe burn is the product of ultraviolet radiation, not the heat of the sun. There is also a host of bacteria and other microscopic life forms which can be killed or controlled by short-wave ultraviolet rays.

The *long-wave* rays, or "black light," as they are sometimes called, are completely harmless. These rays are nearest our visible-light spectrum and can produce reactions in various organic and manufactured materials that the short-wave rays cannot. Similarly, there are many materials, particularly selected minerals, that will not react to long-wave energy, yet they are truly spectacular under short-wave energy. Consequently, collectors of luminescent minerals really need two light sources—one for short frequencies and one for long frequencies—or a lamp equipped with filters to produce both.

Types of Luminescence

What produces the remarkable colors? A basic and technically oversimplified explanation might be as follows.

All minerals are composed of atoms, which in turn are composed of smaller particles called protons, neutrons, and electrons, the latter of which revolve in orbits around the nucleus, or center, of the atom. There is no such thing as solid rock. All matter is, in fact, made up of moving particles. When certain minerals from special localities are exposed to ultraviolet light, the normal pattern of atomic-particle structure is disturbed. In other words, the ultraviolet radiation throws the particles out of normal alignment: they are knocked out of orbit in varying degrees. During this displacement the particles try to realign themselves and, in so doing, give off energy of different wavelengths that we perceive as light of various colors.

While the specimen is exposed to ultraviolet light and turns a color other than its natural color, it shows the property of *fluorescence*. Most fluorescent materials instantly return to their normal appearance once the ultraviolet light is turned off; however, if the particles take a longer time to reorient themselves after the light is turned off, the specimen may continue to glow for a few seconds to several hours. This is the phenomenon called *phosphorescence*, and it is encountered much more infrequently than fluorescence. Only a small percentage of fluorescing minerals will phosphoresce, but strangely, there are some localities that produce specimens that fluoresce weakly, yet exhibit strong phosphorescence! This is dramatically demonstrated by attractive brown, translucent (by normal light) cubes of fluorite from the well-known locality of Clay Center, Ohio. Fluorite from many localities fluoresces well.

However, the Clay Center fluorite yields little reaction in terms of fluorescence, yet shows a very strong pale yellowish phosphorescence.

Some specimens show fascinating growth phantoms under ultraviolet light. For example, some of the translucent cubes of colorless fluorite from the mines around Rosiclare and Cave-in-Rock, Illinois show a beautiful blue-green cube in the center of the crystal under short-wave light. A similar phenomenon is found in some of the fluorites from Clay Center, Ohio, but in this case the internal zoned phantoms are of a pleasing creamy-yellow color. Many of the fine blue-green barite crystals from Chalk Bluffs, Weld County, Colorado have inclusions that show up under short-wave light as intensive apple-green flecks but are totally indiscernible under normal light. Fluorescing phantoms are also frequently found in selenite crystals.

Lamps or instruments generating ultraviolet frequencies are manufactured by various firms throughout the world. There are numerous mineralogical, chemical, germicidal, medical, and other commercial uses for ultraviolet light and for the luminous materials that react to this radiation.

There really is no danger in using ultraviolet lamps, provided normal caution is exercised. As indicated earlier, long waves are totally harmless; however, there are a couple of practical precautions for using short-wave lamps. *Never* look directly into the lamp. Since short waves can give you a superb sunburn in the out-of-doors, the lamp can sunburn your eyes. Consequently, if a short-wave lamp is to be used for any period of time, ordinary reading glasses or some type of plastic goggles should be used for protection. Sometimes a peculiar odor is noted when using ultraviolet lights—this is simply ozone and has no immediate harmful effects, although it is considered an atmospheric pollutant.

Lamp manufacturers produce units for both AC and DC use—in other words, lamps that may be plugged into regular power outlets and lamps that are completely portable. The portable units are popular with collectors who enjoy prowling around old mine dumps or rock deposits at night searching for fine specimens. A word or two of caution: obviously, exploring an old mining area at night can be dangerous in terms of falling into an open shaft. It is best to know what the terrain looks like in daylight first before night explorations are attempted. Another interesting fact is that many organic life forms may fluoresce intensely under ultraviolet light. Scorpions fluoresce a brilliant yellow-green, so try to identify the fluorescing object before you pick it up!

At this point it should be noted that the phenomena of fluorescence and phosphorescence can be produced in certain minerals by sources other than ultraviolet radiation. For example, materials may react to exposure to x rays, high electrical charges in a vacuum tube, or a variety of radiation types generated by atomic reactors. Gemstones may be subjected to such irradiation, which can cause certain diamonds to turn yellow, green, or even orange. The color of a wide variety of transparent crystallized minerals, from quartz to topaz, may be altered drastically by atomic radiation. Most of the time the changes in color are permanent, but some fading takes place in a few days or months.

There are several fascinating variables involved in this phenomenon. The most conspicuous lies in the fact that a specific mineral from one locality, for example, calcite or willemite, may fluoresce strongly—while a specimen from another locality may be completely "cold." As a matter of fact, there are relatively few localities that produce spectacular fluorescent specimens of calcite, even though calcite is an abundant worldwide mineral. Apparently a very delicate balance exists between the mineral (calcite) and some impurity that acts as a catalyst, triggering the fluorescence or phosphorescence. Much more research is needed in this field.

A good example can be found in the calcite found in reasonable abundance at various old zinc mines in the vicinity of Franklin and Sterling Hill, New Jersey. Perhaps the world's most spectacular fluorescent-mineral specimens have come from this area, with single specimens displaying intense red, green, yellow, and pink fluorescence. In the case of the calcite, it fluoresces in various shades of deep red to pale pink, apparently due to the inclusion of from 1 to 5 percent manganese as an impurity. Specimens containing from 3 to 3½ percent manganese exhibit the most intense cerise-red fluorescence. If the specimen contains *less* than 1 percent or *more* than 5 percent manganese, there is no fluorescence! Consequently, not all of what *appears* to be identical material from a specific locality will fluoresce.

Perhaps the most interesting mineralogical use of fluorescence occurs with the mining of the metal tungsten, of which one of the prime ores is scheelite. Fortunately, scheelite seemingly has the worldwide attribute of fluorescing a strong blue-white or yellow-white color. Scheelite in an ore specimen can be absolutely indistinguishable to the eye under normal light. However, expose the specimen to ultraviolet light and the valuable scheelite areas burst into color. Obviously, ultraviolet light is very useful in locating ore bodies containing this tungsten material. Similarly, ultraviolet is employed in tungsten mines to indicate clearly ore-body concentrations for subsequent removal.

Incidentally, impurities play a part in the different shades of fluorescence observed in scheelite. Scheelite is calcium tungstate and is commonly found with small amounts of molybdenum associated with tungsten. If there is little or no molybdenum involved, the scheelite will fluoresce a brilliant blue-white. If molybdenum is present, the scheelite will fluoresce yellow-white, with a deeper color indicating a higher molybdenum content.

There are several other minerals of commercial value that may show fluorescence for identification purposes. Willemite, hydrozincite, zircon, some sphalerite, and several of the uranium-ore minerals such as autunite are examples. However, intriguing variables are always present. The massive willemite from Franklin, New Jersey fluoresces an intense, brilliant green, while the same mineral occurring magnificently well crystallized at Tsumeb, South West Africa exhibits absolutely no fluorescence. The same applies to fine crystallized-willemite specimens from the famed but now inoperative Mammoth-St. Anthony mine, Tiger, Pinal County, Arizona.

In terms of the various uranium-bearing minerals, there is a particularly interesting and not altogether understood aspect of one type of deposit. In California, Arizona, New Mexico, and especially in Utah, heavy concentrations of carnotite and other uranium-bearing minerals within agate—from very selective localities—will cause the specimen to fluoresce a pale, ghostly green. Many, but not all, waterworn agates from Wyoming, known as Sweetwater Moss Agates, will fluoresce this fine color. Agate from many localities in the western United States has been found to exhibit a pale lime green to yellowish green color with a weak to modest intensity.

Some minerals, such as calcite, aragonite, willemite, halite, barite, common opal, wernerite, selenite, witherite, adamite, hydrozincite, apatite, fluorite, diamond, and celestite, may be more prone to show fluorescence than others. This simply means that selected specimens from certain localities may exhibit the phenomenon, and wherever they are found they *might* fluoresce. Most mining districts contain some sort of material that will show some degree of fluorescence, even if it is only a paper-thin coating of aragonite on a rock—it might fluoresce or phosphoresce. At selected localities fluorescing specimens may be abundant, and at others scarce. There is no rule of thumb.

There are two other interesting aspects of observable mineral luminescence worth mentioning. Some minerals, notably selected calcite and fluorite specimens from certain localities, will emit a slight glow when low heat is applied to the specimen. This is called *thermoluminescence*, and the atomic-particle orientation of the mineral is apparently disturbed by the heat factor. Note that this does not mean that the specimen is heated until it is "red hot" but involves only a modest heat application.

Another unusual visible phenomenon is shown by certain specimens of corundum and sphalerite. This is *triboluminescence* and is characterized by an internal flashing, or sparklike emission, when the specimen is scratched by a sharp metal point. It may also be observed when a quartz crystal or agate is cut by a diamond saw. In this case the sudden, abrupt application of pressure apparently disorients the atomic structure by building up electron charges on either side of the contact point which subsequently arc across this gap, producing light. The phenomena of both thermo- and triboluminescence are best observed in maximum darkness, since the effects are weak or fleeting.

Chapter 8

CONCRETIONS AND SOME "FOSSILS" THAT AREN'T

Museum curators, advanced collectors, instructors, and specimen dealers are frequently faced with the question, "What kind of a fossil is this?" Such inquiries occur on field trips, after a mineral-society talk, during classes in mineralogy or palentology, at a dealer's shop, or at the home of a collector. Many times they have to be told, "Sorry. It isn't a fossil—it just looks like one." The novice may reply with a cold stare, usually followed by a somewhat incredulous, "How do you know it isn't?" or, more aggressive, "It sure as heck is. Just take a closer look—right here where all the ferns show!"

This opens up the intriguing door to another interesting facet of geology (in this case a combination of paleontology and mineralogy), which deals with specimens that look like fossils but are *not*. Fossils should be considered as any type of evidence of prehistoric life, whether they are permineralized specimens, casts, molds, impressions, or the product of simple preservation. Many true fossils are, indeed, difficult to identify, a problem that is complicated by a number of mineralogical mimics, look-alikes, that are real foolers. Various mineral forms may display an amazing superficial resemblance to bones, plants, shells, wood, or coral. For lack of a better term, such mimics might be called "false fossils"; a great number of perfectly natural objects occur in rock strata, ore bodies, or as products of erosion that resemble ancient life forms to an astonishing degree.

Concretions

Concretions may or may not be associated with fossils and may or may not look like fossils. Concretions are fairly common and often far more interesting than the simple rock they may appear to be. They are a type of geologic oddity found in sedimentary-rock strata throughout the world.

In geology and mineralogy a *concretion* is one of the frequently encountered physical forms in which various minerals may occur. Essentially concretions are nodular, spherical, or oddly shaped rounded masses representing the deposition of mineral materials around some sort of original nucleus. Sometimes the deposit will show distinct, successive shell-like layers of mineral matter (like the annular rings of a tree) when the specimen is broken or cut open. Concretions may vary in size from that of a pea to monsters over 15′ in diameter. Many huge examples have been found in Wyoming, Montana, Texas, and Utah.

From the standpoint of the mineralogist, many different minerals may show concretionary form—a spherical, rounded mass that originally grew from a

Figure 8-1. No, it's not a fossil ground squirrel! This is a good example of how *concretions* often resemble life forms such as animals, bones, wood, or shells or even man-made artifacts. These sedimentary formations are very common throughout the world and may vary from the size of a pea to as large as an automobile.

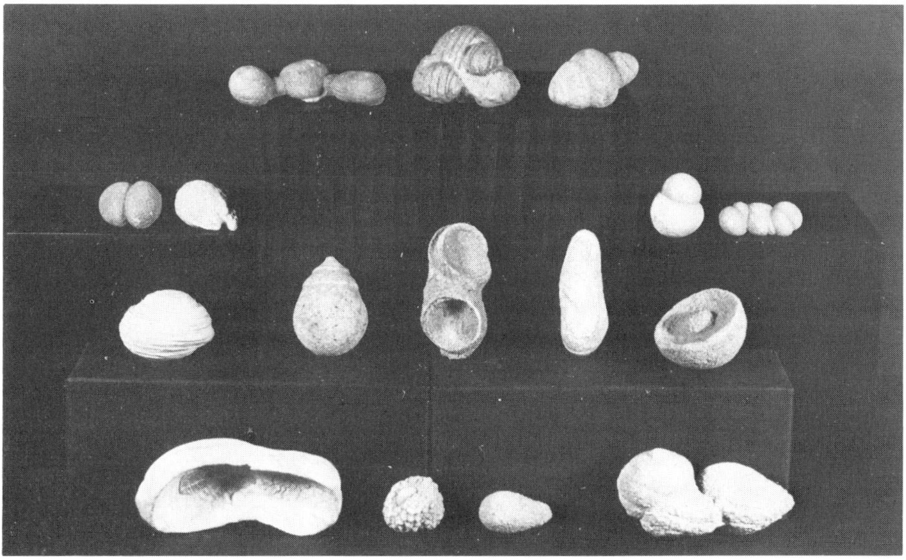

Figure 8-2. A sampling of the bizarre shapes common to concretion beds around the Salton Sea in California. These are all small, but the variety of forms extends into large specimens as well. None of these contains fossils.

nucleus. Calcite, barite, aragonite, quartz, pyrite, marcasite, limonite, vivianite, and hematite are species often found as concretions. One notable locality for fine green variscite of both gem- and museum-display quality is in Utah. The specimens occur as concretions of this unusual phosphate mineral in a hard limestone.

The precise mechanics of the growth of concretions is not thoroughly understood. However, prior to exposure to weathering, all concretions are an integral part of some type of host-rock stratum. Normally, concretions are significantly harder than the surrounding rock and consequently erode from the host formation as isolated spherical or roundish "rocks." Evidence seems to indicate that some concretions are generated within an existing rock formation as some nucleus attracts mineral solutions to it and actually pushes material around the growing form aside. Other evidence suggests that concretions form in existing cavities in the rock, perhaps caused by gas bubbles, and gradually fill the cavity with layer upon layer of minerals. Such formations, however, are probably closer to a nodule form than to any other classification. The difference between a nodule and a concretion is often borderline at best. There are certain types that definitely overlap, and authorities may use the terms somewhat interchangeably in certain cases.

Figure 8-3. Selected localities throughout the world have furnished concretions enclosing fine fossils—crabs, teeth, shells, leaf and frond imprints, or skeletal materials. This fine ammonite formed the original nucleus for a large concretion found in southern Utah. The concretion matrix has been broken away to expose the fossil. (Specimen courtesy of Rosemary Luxton.)

Many geology and mineralogy books often refer to a type of specimen called a septarium nodule. It is simply a specialized type of concretion that shows a peculiar intersecting series of veins, or divisions, on its surface. Sometimes these divisions extend throughout the mass of the concretion. Septarium concretions have not yet been fully explained, but in all probability the odd cross-veining is produced by contraction or expansion of the original formation during growth, causing it to fracture or crack. The cracks in the spherical mass fill in with a harder mineral and produce the divided checkerboard-surface effect.

Perhaps the most interesting and oddest concretions are those that apparently began to form around a nucleus in an ancient lake, pond, swamp, delta, or sea bottom. Today we find this type as hard sandstone, siltstone, or claystone formations. Usually the mineral acting as the cementing agent is calcite. The original nucleus that founded the concretion may have been a fern frond, insect, shell, bone fragment, arthropod, tree or shrub branch, leaf, fish, tooth, or scale that settled to the muddy bottom and gradually accumulated mineral materials around it. Eventually these mud balls hardened, were assimilated into the rock strata in which they formed, and became concretions.

Foreign objects continue to attract silt and mineral materials in a watery environment even today. Bottles, dishes, beer cans, and tires drifting into a mud-flat area will soon collect mud and quickly become completely encased. Perhaps millions of years from now some scientist will crack open a large, round, flat concretion and be faced with a tire imprint—then the fun will begin trying to identify it.

In selected localities concretions may be the source of very fine fossil specimens. The famed concretions of the coal deposits around Mazon Creek, Illinois reveal perfect imprints of a wide variety of ancient plants when split open. Another locality near Mt. Carmel, Utah has produced concretions containing beautifully preserved coiled ammonites. Similar deposits in South Dakota and Texas have yielded a variety of fossils, and a locality in Brazil has produced hundreds of 1′ to 2′ oblong concretions, each containing a well-preserved skeleton of a very primitive type of fish.

One of the world's most unique and fascinating deposits of fossil-bearing concretions is found in the Mojave Desert a few miles north of the town of Yermo, San Bernardino County, California. Specifically, there are three localities within a short distance of each other, centered around the Mule Canyon area of the colorful Calico Mountains. Here in the bed of an ancient Miocene lake or series of ponds in the Barstow formation are found at least three levels of strata containing many small, very hard, calcareous concretions varying in size from around ½″ to 2″. A great number of these little concretions contain the exceptionally well-preserved remains of arthropods and insects, in many respects surpassing the known occurrences of such life forms in amber (fossil resin). Over twenty different species of nonmarine arthropods or insects have been found, including dragonflies, beetles, midges, mites, fairy shrimp, spiders, and leafhoppers. A high proportion of the fossils are in larval stages, but adult forms are also present, and they represent both aquatic and nonaquatic fauna.

But the external and the internal structures of many of the specimens are amazingly well preserved, and fairly complete specimens as well as individual silicified parts are found. The fossils are quite remarkable—in numerous cases it is apparent that mineralized replacement took place soon after death or before decay took place, since internal organs are found in a perfect state of

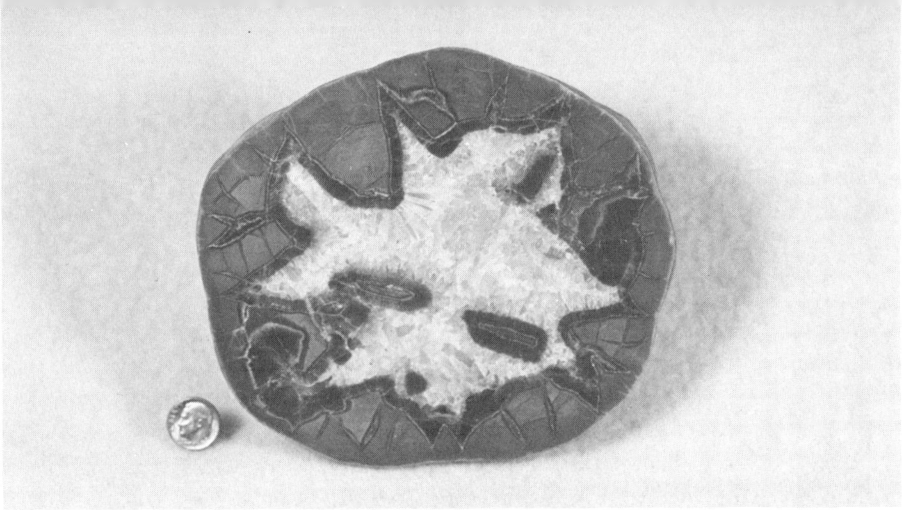

Figure 8-4. Septarium concretions from near Mt. Carmel, Utah are filled with bright yellow calcite phasing into rich brown against the concretion matrix. These specimens have been mined commercially for years. Infrequently they contain fine ammonite fossils.

Figure 8-5. These small and very unusual concretions come from beds in the Calico Mountains of the Mojave Desert. Most contain well-silicified fossil remains of insects and arthropods. The center specimen is probably a larval stage of an ancient dragonfly. The concretions are petroliferous—a box of them will yield a petroleum odor.

preservation.

Obtaining the fossils, however, is not as easy as it might sound. The little nodules, or concretions, are carefully dissolved in diluted formic acid, and the insoluble fossil remains are found in the acid residues. Many of the most perfect fossils are best viewed through a microscope. There are some half dozen minerals involved as replacement agents, but the most abundant is quartz, which replaces the life forms in the utmost detail. Other replacement minerals are celestite, gypsum, calcite, bassanite, and analcite. Often the life forms are replaced by an unidentified dark brown organic compound. Infrequently, very small chips of well-silicified bone are found associated with the nodule or the concretion stratum.

Another interesting aspect of these Calico Mountain fossil-bearing concretions is that they are *petroliferous;* i.e., many of the nodules will yield a droplet of petroleum when dissolved in acid, which may be seen as an oily scum on the surface of the acid solution. A box full of these little concretions yields a very marked odor of petroleum.

The foregoing illustrations are not meant to imply that all concretions are started by life forms. Far from it! Most areas that produce concretions show absolutely nothing that is now discernible as a life form. These may have originated with a fragment of rock; a life form that completely disintegrated, leaving no trace; or perhaps a purely chemical original nucleus.

Once exposed to the elements, most concretions seem to have a relatively meager life and may disintegrate in a short time. However, a few localities, notably the Book Cliffs near Grand Junction, Colorado and Elk Creek, South Dakota, have produced large, exceptionally *hard* concretions, the centers of which are geodal. These have been the source of spectacular museum or collector specimens of water-clear or lustrous transparent yellow-brown crystals of barite. Similar very hard concretions from coal-strip-mining areas near Knoxville, Iowa have furnished handsome specimens of tiny brown, iridescent calcite crystals encrusted with beautifully developed "bursts" of snow-white acicular calcite crystals.

Another type of concretion that often produces amazing fossil look-alikes, usually distinguishable only by an advanced amateur paleontologist, are similar to fossil sponge and algae colonies. Cutting and polishing the specimen will usually reveal the details of the specimen structure.

Certain localities have furnished sandstone concretions in a variety of shapes and sizes that might fool the inexperienced eye. One particular locality (now almost exhausted) is the area surrounding Truckhaven, Imperial County, California. Here concretions could be found resembling tools, grapes, lemons, peanuts, acorns, dumbbells, tops, balls, and other objects. A bewildering and bizarre assortment of sizes and shapes could be easily obtained by walking the various washes of the area and digging in the soft beds of eroding rock where the concretion forms were noted. As is the case in most concretion beds, there is nothing inside—they are solid sandstone.

In the 1930s, along with cactus gardens, southern Californians indulged in the fad of displaying concretions in their gardens. At this time there were several extensive concretion beds in various sections of southern California, particularly in southern Imperial County, eastern San Diego County, and Orange and Los Angeles Counties. These localities are now virtually exhausted, but they produced literally thousands of concretions, ranging in size from a quarter to specimens as large as a man.

The proverbial rock or cactus shop of those days often featured concretions. There seemed to be an anthropomorphic concretion to mimic everything and anything, whether it was a silhouette of W. C. Fields, an automobile, cat, turtle, elephant, or Gibson Girl! Again, the degree of freakish simulated realism found in many concretions makes them look almost as if they had been carved or cast as lawn ornaments!

No discussion of concretions would be complete if mention were not made of a real puzzler, a truly unique type of concretion that apparently remains unexplained—*sand spikes*. The original locality is believed to have been discovered by Mr. H. W. Pierce of Laguna Beach, who amassed a fabulous collection of the fascinating formations and used them as trade items to assemble a very fine mineral collection. Mr. Pierce also made generous donations of these amazing formations to museums throughout the world. A few years ago, my wife and I were visiting the superb mineral collection displayed at the Museum of Natural History, Lyon, France and were delighted to see a case devoted to sand spikes, courtesy of Mr. Pierce!

Briefly, the locality consisted of a series of low, sandy hillocks and banks near the Mexican-American border close to Mt. Signal, Imperial County, California. When they were first discovered, many of these unique concretions were weathering out on the surface of the ground. It became apparent, however, that there actually were beds of these strange concretions from 3′ to 8′ underground!

Figure 8-6. These strange concretions, called *sand spikes*, came from a now-depleted locality in California near the Mexican border. These provocative formations of different sizes were found in beds of sand, and they all point to the west!

The following may be almost too much to believe, but, having collected at the locality before it was totally decimated, I can assure you it is quite accurate. These concretions consist of a ball-like end coupled with a tapering spikelike formation. In uncovering a bed of spikes over 95 percent of the formations were found *pointing west*. Spikes of similar types or format seemingly occurred in the same bed or within lenslike concentrations of the sand-spike formations.

They range in size from 1½" to over 3' in length and occur in a wide variety of shapes—perfectly round with stubby or long, tapering spikes or knobby with rough spikes. The spikes were usually quite damp and broke easily when uncovered, so they were dried in the sun before being handled. They are composed of absolutely nothing but the identical sand forming the soft little hillocks and banks in which they occur—solid sandstone with no fossil material inside. The mineral cementing the grains together is apparently calcite.

In addition to the fact that they point west, which is really intriguing, it should be noted that some were found in beds of very fine, layered sand of different colors. The different layers in the loose sand beds were found to continue directly through the sand-spike formation itself. Dark layers of iron-bearing sand show in some of the spikes and extend into the soft enclosing sand beds in the same precise stratification as shown in the concretions!

What are they? No one really knows—other than they obviously fall under the heading of concretions by virtue of their method of occurrence and basic mineralogical composition. The long points, westerly oriented, are the most provocative aspect of this truly unique locality. There is no surrounding evidence in the area to indicate any realistic fossil origin. However, there have been some truly far-out suggestions offered. Some say that they are fossils of an ancient type of bulbous seaweed that drifted heavy-end eastward on some ancient shoreline. Others have suggested that they have somehow been affected by an unknown magnetic force that oriented their westerly growth. It has also been suggested that they are some type of fulgerite, which is positively incorrect, and not by any stretch of the imagination are they prehistoric man-made artifacts.

In any event, it is always interesting to note that there are, indeed, geological formations that still remain surrounded by an aura of mystery. Incidentally, this locality has been totally exhausted by pick, shovel, and most unfortunately, bulldozer. The latter technique was very unsuccessful, essentially managing only to crush many hundreds of fine specimens.

False Fossils

Next we should review some fascinating concretions and other related forms that comprise an admirable troop of mime performers. The curtain now rises on the stage of *false fossils*.

Perhaps the most common of all false fossils are those that resemble tiny fern fronds or colonies of algae. Unfortunately, many a souvenir-type shop has pawned these specimens off on unsuspecting tourists as the real thing. Fossil or not, these specimens, called dendrites, are interesting. A *dendrite* is a comparatively common microthin arborescent coating of one of the manganese minerals, called pyrolusite: black manganese dioxide. The specimens most frequently mistaken for fossils are delicate branching, fernlike encrustations occurring in thin, hard bedded shales and limestones.

There are countless localities within the United States that have produced thousands of rock specimens showing various degrees of dendrite growth.

Figure 8-7. Although the resemblance is strong, this is definitely *not* a fossil-fern imprint. It is a *dendrite* of the manganese mineral pyrolusite. Dendrites are very common. This specimen is on shale from Hamilton, Illinois.

Often the dendrites seem to have been deposited along a fracture in the rock, and when the specimen is viewed, it closely resembles a scene of trees along a river, the skyline, or an undersea garden. Dendrites may be very small or quite large: magnificent specimens quite suitable for framing have come from Germany.

It is also possible to find dendrites associated with true fossils. Infrequently they have been found in the same shale deposits of Wyoming that have produced thousands of fine fossil-fish imprints.

Dendrites are found frequently as inclusions in agates, and here again they simply mimic life forms and are in no way related to true fossils. Dendritic agates can be very beautiful and, when cut and polished, may produce handsome jewelry. Many unsuspecting tourists in Brazil, India, Wyoming, or Montana have purchased a "moss" or "fern" agate ring or pendant believing that the attractive little "plants" in the stone are ancient fossils. Not so.

While we are on the subject of fernlike false fossils, mention should be made of the minerals, such as rutile, kaolinite, pyrolusite, goethite, chlorite, and pyrite, that often occur as inclusions in clear quartz crystals as well as in the amorphous forms of quartz such as agate and/or chalcedony. Again, multiple mineral growths appearing *inside* another may often look like branching vegetation, fern fronds, or even flowers. The so-called moss agates or plume agates are simply agates with various thick dendritic or plumose inclusions, often very colorful but definitely not fossils. Note, however, that quartz (silica) is very frequently the *prime* mineral that replaces a former life form—and creates a fossil. Petrified wood is a typical example, since most of the time the permineralizing agent is silica—quartz.

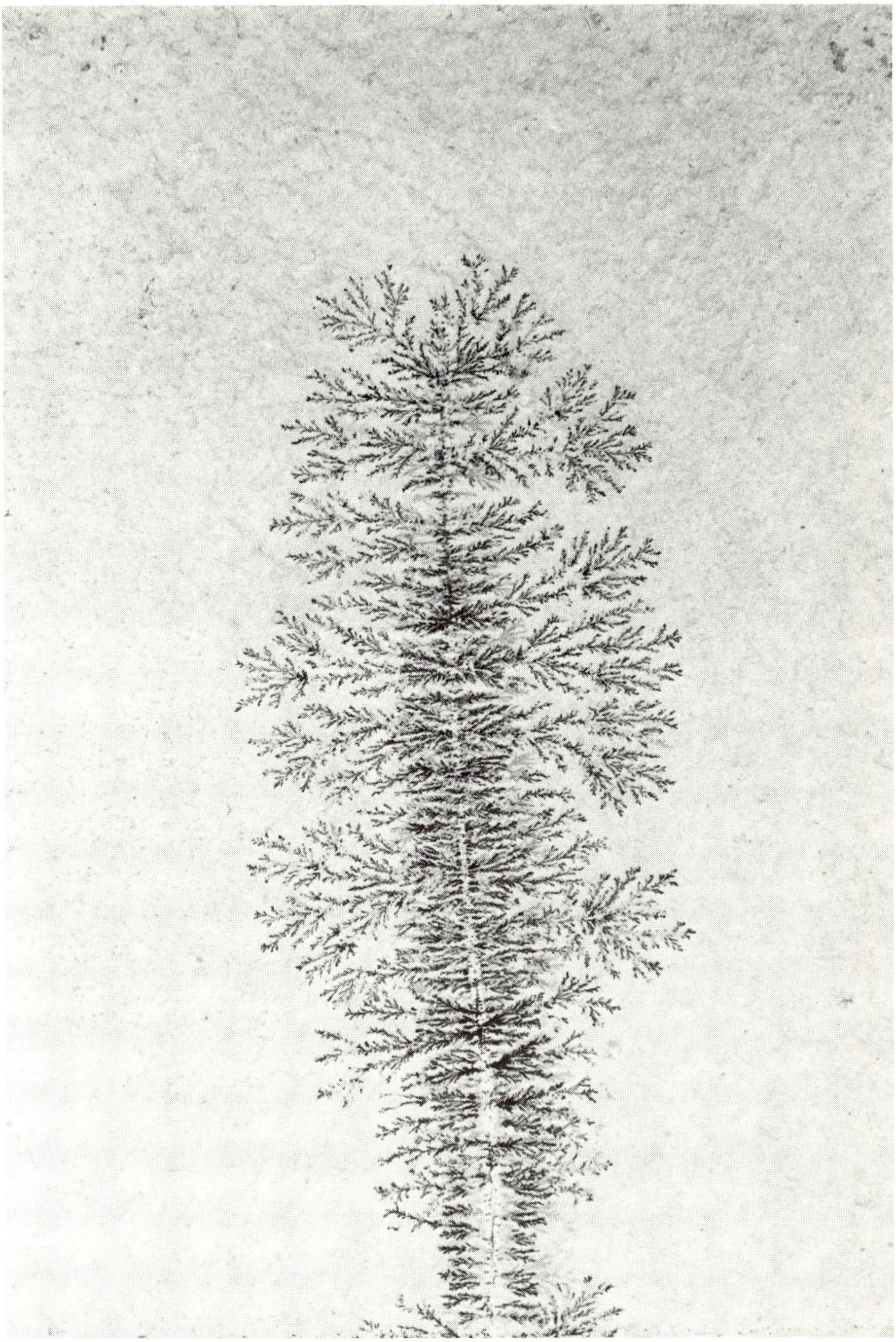

Figure 8-8. Dendrites of black pyrolusite often mimic fossils to an astonishing degree. Some of the finest specimens occur in a very hard shale at Solenhofen, Germany. The shale is split and the dendrite area sawed into sizes appropriate for collections, museum display, or framing. (Specimen courtesy of Kristalle.)

Another *extremely* common fossil mimic is the mineral, or rock, chert. Chert frequently occurs in different-colored layered specimens, often concentrically banded much like the annular rings of a tree. Consequently, the ability of chert to mimic petrified wood or even bone often constitutes a first-class performance!

Chert is abundant, and a typical pebble could easily pass for a waterworn cobble of petrified wood. Telling the difference between chert and real petrified wood is more easily demonstrated than written about! When in doubt, collect the specimen and inquire of someone knowledgeable. Chert will never show any cell structure, but some true petrified wood doesn't either. Chert has probably confused more collectors than any other material (other than dendrites) involved in the false-fossil field.

Sometimes agate-nodule beds produce "lifelike" specimens. It is quite possible for gas bubbles in ancient lavas and even in thick ash deposits to assume shapes resembling pears, lemons, figs, or even the individual septum from some type of citrus.

Another source of fossil-appearing specimens is associated with differential erosion. Simply stated, this means that certain rock or mineral materials are either harder or softer than the material deposited on either side of them. There is a tremendous range of relative hardness in rocks and minerals. Some, such as diatomaceous earth or talc, are exceptionally soft, while other materials, such as diamond and basalt, are exceptionally hard. Between these two extremes there are hundreds of variables. A simile for our purposes might be that the covers on a book are harder than the pages in between. Likewise, butter is softer than the carton in which it is enclosed.

Different layers of rocks or minerals are frequently deposited in sequence but vary greatly in their respective hardness. Consequently, this produces a marked difference in how a fragment of that material or stratum will be affected when it is subjected to the forces of snow, frost, running water, and wind. The harder layers will be more resistant to erosion than the softer, and this can produce weird and interesting forms.

On a grand scale examples may be found in Goblin Valley of Capitol Reef National Park, the narrows of the Grand Canyon in Yellowstone National Park, Arches National Park, Bryce Canyon National Park, and Cedar Breaks National Monument. However, the most intriguing of all grand-scale scenic areas featuring differential erosion consists of the menagerie of "cartoons" in stone found in Chiricahua National Monument, Arizona. In all cases the scenic attractions of the mentioned areas are the product of erosion working its telling effects over eons of years on hard, intermediate, and soft interbedded layers of rock materials.

Small, hand-size specimens are found frequently that are also the product of this perfectly natural process of differential erosion and may resemble fossils to a remarkable degree. Rocks that break out of strata containing two hard layers with a softer layer in between may weather to look very much like bones. Other waterworn rocks may roughly resemble crinoid buds, seashells, or wood. Close inspection usually reveals that the similarity is superficial, particularly if a freshly broken surface is examined. However, there are many cases where real, true fossils may be very difficult to identify as the real thing. Most of these instances are resolved after identifying the conditions in which the specimens are found.

Figure 8-9. A waterworn pebble of chert, a close relative of flint and a microcrystalline granular form of quartz. Chert is very common, occurs in a wide variety of colors, and is often layered in structure. As this specimen shows, chert may be easily mistaken for petrified wood.

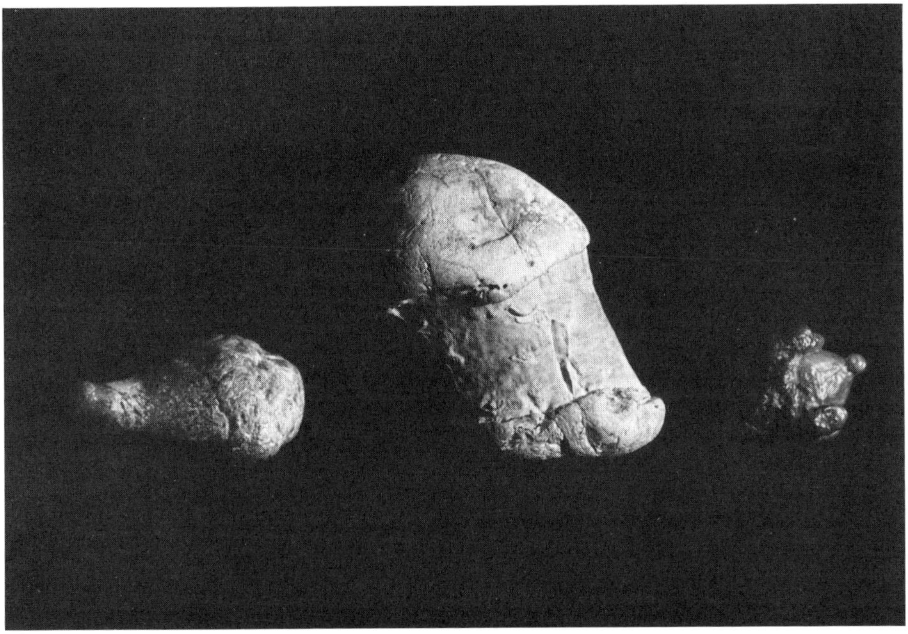

Figure 8-10. Fossils? No, these are simply rocks containing sections of different hardness that have eroded in an uneven manner. They are typical of innumerable specimens found in the field that may resemble life forms but are merely the product of differential erosion.

For many years staff members of natural-history and geology museums as well as some professional dealers have been expert in making casts, reproductions, or models of outstanding fossil specimens for exhibit, instructional, and general sales purposes. The technology involved in the handling of various plaster and plastic materials has almost become an art form! The professional museum preparator can reproduce a small bone, a huge skull, or a plate of crinoid stems with absolute perfection; you could not tell the reproduction from the real thing without very close inspection.

There are several companies that publish detailed catalogs of paleontological reproductions and models, as well as offering a wide variety of real fossils for sale. Again, the flawless accuracy in terms of appearance, detail, and even weight

Figure 8-11. All of the "fossils" in this exhibit are man-made reproductions. Modern technology can duplicate an authentic specimen so closely that only an experienced collector can tell the difference.

is remarkable. Some of the reproductions are so good that they will withstand microscopic examination. Usually such items are sold to museums or to colleges and universities for exhibit purposes or as instructional aids. They are a valuable resource, since reproductions make available excellent examples of very rare fossils for student study. Many of these items find their way into museum shops throughout the world, and a fine reproduction of a dinosaur egg, a large shark's tooth, a trilobite, an upper canine from a saber-toothed cat, or a tyrannosaurus tooth makes an excellent souvenir of a memorable visit. It also makes a good decor item and is a great conversation piece!

Museum shops and reliable dealers always sell this merchandise clearly marked as a model or reproduction. Firms selling sets of fossils as instructional aids clearly label the specimens as replicas. However, in recent years man-made fossils have appeared in many other facets of the consumer market, such as souvenir and curio shops or gem and mineral shows. There is little doubt that many people have purchased vinyl trilobites without having been supplied with correct data. Instead of a true 600-million-year-old fossil, they now own a 1- or 2-year-old specimen of plastic.

Other real mind-bogglers in geology are the amazing look-alike specimens called *cone-in-cone* formations. Many novices and even experienced collectors have been thoroughly deceived in identifying these specimens as fossilized wood. The resemblance is astonishing.

When you handle and examine a good example of a cone-in-cone formation, your first thought will be that it must be petrified wood; if not, it simply must be some other type of fossil. Specimens can look exactly like wood, even fossilized knots, and in some specimens it appears as if some prehistoric animal actually gnawed on the wood. Sorry—wrong on all counts, although they would be logical guesses.

Cone-in-cone formations are superb examples of false fossils and constitute a well-established geological oddity. These peculiar formations vary considerably in size: they may be 1" in length or over 4' long.

Just how they are formed is not too well understood. However, they definitely are the product of intense pressure on extremely fine silts or muds, squeezing the soft material upward through a crack or joint. This is probably coupled with a series of mineralization processes as one cone becomes coated perhaps with calcite, allowing another cone to be forced up under it, and so on. Sometimes small crystals of pyrite or pseudomorphs of limonite after pyrite are found embedded within the cone-in-cone structure.

For some strange reason, the physics involved produces plates, or strata, containing individual lenslike nests of these conelike formations. They have been found sparingly at various localities throughout the world in sedimentary strata, especially in Montana, Oklahoma, Kansas, and Austria.

Among the numerous terms used in physical mineralogy to describe the various forms in which a mineral may occur, there are three that suggest life: *coralloidal* (corallike), *arborescent* (treelike), and *vermiform* (wormlike). The flos-ferri type of aragonite resembles coral growths; native copper is frequently found in arborescent sheets that look like a cluster of flattened shrub branches; both aragonite and calcite can occur in strange curved vermiform aggregates. Of course, these are not fossils but merely descriptive terms that indicate the basic shape or physical appearance that certain minerals may assume.

Figure 8-12. One of the most intriguing of all fossil imitators are *cone-in-cone* formations. The resemblance to fossil wood is often incredible. The specimen in the middle is petrified wood; the others are cone-in-cone formations. They are composed of extremely fine sedimentary material that has been subjected to unusual pressure conditions.

Sandstone Landscapes

There remains one other somewhat related oddity that is indeed neither a fossil, real or false, nor any type of concretion. Nevertheless, since sandstone is involved, the specimens are closest to home in this chapter. This is the phenomenon of *sandstone landscapes*, particularly those from Utah.

From this introduction the reader might expect a discourse on Utah's famed scenic wonders such as Arches, Canyonlands, Zion, Bryce Canyon, and Capitol Reef National Parks or perhaps Cedar Breaks, Natural Bridges, or Rainbow Bridge National Monuments. Here the prime scenic elements are colorful eroded shale, limestone, or sandstone formations. Although much of Utah's general landscape is indeed intriguing, so is another "solid" aspect of Utah's geology. These are the unique and limited deposits of "picture rock," or "picture stone," an unusual layered sandstone found in commercial deposits at three Utah localities. These materials are absolutely not fossils.

Did you ever hear of anyone mining scenery? Well it is done near Kanab and St. George, Utah on a professional and commercial scale, with the most workable deposits under private ownership and claim just like a gold or silver mine.

Imagine a cut slice of sandstone that shows gentle rolling hills, a river in the foreground, mountains in the distance, and a few puffy clouds in the sky, or perhaps towering thunderheads! Such scenes as these are characteristic of the amazing realistic, perfectly natural pictures in sandstone rock. Specimens often resemble fine, highly stylized paintings, yet they are positively *natural*. They can be so realistic that they might be considered fakes.

There are two well-established major localities for this unusual sandstone in southwestern Utah, and each produces distinctive specimens. "Picture rock" is the local name given to material found at a site neat Kanab, Kane County. Here the sandstone is fairly coarse, and the colors involved in the scenes are various marked shades of rust, tan, brown, buff, and violet. The second important locality is near St. George, Washington County, and it produces specimens locally called "picture stone." This material is a very fine-grained sandstone in which the various scenes appear in delicate and subtle shades of brown, tan, fawn, and mauve.

These two localities have been worked since around 1947, and obtaining quality material became somewhat difficult, since the deposits are not extensive. However, in 1973 a third high-quality locality was found near the original Kanab site. Strangely, this newer locality produces very fine material that shows color characteristics of both the older Kanab material and the site near St. George.

Figure 8-13. This is a typical landscape scene in a cut slab of fine-grained sandstone from Kanab, one of southern Utah's three localities for this odd material. These specimens are very effective mounted and framed as pictures. The colors here are buff, brown, salmon, red, and black. These interesting specimens are in no way related to fossils.

The rock itself was deposited sometime during the Triassic Period, some 150 million years ago. At that time the region was low-lying country composed mainly of swamps, lagoons, and sluggish streams that deposited hundreds of feet of silt, later to become sandstone. The beautiful picture rock, or picture stone, is actually a very thin, fine-grained sandstone layer of what geologists call the Shinarump Formation, and it lies between the Moenkopi and Chinle sandstones.

The origin of the various colors producing the fascinating scenes lies in the precipitation and deposition of various iron and manganese oxides diffused from solutions permeating the sandstone rock. When the layers of sandstone are cut at right angles to their original bedding position, they reveal an almost endless series of rhythmic, undulating scenic patterns.

The deposits are privately operated as mining claims, and the amount of material available is definitely limited. The rock is quarried by hand and tractor. Boulders are split with large chisels along natural seams in the rock and transported to workshops for further preparation. The rock is cubed and slabbed on a diamond saw to reveal the scenic patterns and then dressed with a piece of softer sandstone to remove the saw marks. The material is fairly soft and easily worked. Oddly enough, hand finishing at this stage has proved most satisfactory, although a variety of mechanical techniques have also been tried.

Today choice slabs of this fascinating, colorful material have found a ready market in glass- or plastic-protected tabletops and award plaques. However, the major market is in slabs that are framed and mounted exactly like a picture. These natural masterpieces seemingly blend in with any decor—office or home. With a fine specimen of framed picture rock or picture stone the owner may rest assured that he has a true original!

Another general area of considerable importance in furnishing a wide variety of fine gem-quality scenic materials is the long-known Succor Creek area around Owyhee Lake in Malheur County, Oregon. There are at least a dozen localities scattered through this area, producing a variety of different multicolored, layered cryptocrystalline forms of opaque quartz. When they are cut and polished, these very hard specimens may show rolling hills, mountains, prairie scenes, seashores, and even towering clouds in a blue sky. These materials are known under several local names, but Succor Creek Picture Rock and Owyhee Picture Jasper are perhaps the best known. The scenic structure is frequently on such a small scale that selected specimens can be worked into very attractive cabochons or flats for pendants and bracelets.

Chapter 9

FORESTS TURNED TO STONE

One of the most intriguing and fascinating facets of paleontology and mineralogy involves the remains of ancient forests now turned to stone. Fossil forests have tremendous appeal to the tourist, neophyte, and experienced paleontologist as well as the mineral collector. They are a prolific oddity.

Fossil forests, or concentrations of fossil-wood specimens, are not rare. Wherever there are ancient sedimentary rocks or rocks of volcanic-debris origin, there is a possibility of fossil wood. There are extensive fossil forests in Siberia; fossil-redwood specimens have been found at numerous sites throughout Europe; and an extensive forest is found in Patagonia, Argentina. The United States is richly endowed with fossil wood, and specimens have been found in all of the continental United States. In many areas, fossil wood is actually abundant, and it may well be the most common of all fossils in terms of

Figure 9-1. The Eden Valley area of Wyoming is a well-known locality almost exclusively devoted to small fossil limbs and branches. The little limbs often show remarkable cell structure, and frequently the limb itself is completely encased in a shell of translucent chalcedony. Specimens can now be collected only by extensive digging.

distribution. There is a reason for this. In retrospect, there have been billions of trees or treelike plants on our planet as prehistoric life developed during the last 600 million years. Trees seemingly lend themselves well to the processes of fossilization and have had innumerable opportunities to become entrapped or encased by materials that later become rock. Once buried, the process of fossilization begins. Entombment may be caused by landslides, encased in mud and silt or showers of volcanic ash and ejecta. Trees may be buried as prone, prostrate logs or entombed in an upright position.

The factors of prehistoric abundance, especially when compared with other life forms, and the variety of conditions providing entombment opportunities have produced dozens of major localities where concentrations of fossil wood are large enough to be called a *fossil forest*.

Well-known fossil forests of major proportions, protected or unprotected, are found in Arizona, Wyoming, Colorado, Oklahoma, California, Oregon, Washington, Idaho, and Nevada. Many of these states, Nevada and Wyoming in particular, have dozens of wood-producing localities within their borders. With few exceptions, most of the wood deposits have been well-visited by hobbyists for years. In some cases, commercial lapidists or dealers have stripped an area of all accessible material. Nevertheless, fossil wood, often of fine lapidary quality, can still be found at numerous localities in the West. There are ample booklets, field-trip guides, and articles in the mineral and lapidary magazines on the subject.

Without a doubt, more fossil wood is sold to tourists than any other rock, mineral, fossil, or gemstone. Many localities furnish beautifully agatized or jasperized material that makes colorful and interesting cabochons for rings, bracelets, and necklaces. The tourist business probably consumes thousands of pounds of this type of wood from the Petrified Forest National Park area in Arizona each year. Collecting specimens in any National Park area is, of course, against regulations. The Arizona wood that is marketed so heavily comes from outside the Park boundaries but from identical strata. Most of these deposits are on private property, and the material is collected or mined commercially. It is every bit as attractive as that within the Park.

There are many hobbyists who specialize in fossil woods. This is another of the earth-science hobbies that can be as casual or as sophisticated as you desire. The study and determination of fossil woods and other plants is called paleobotany, or, more specifically, paleodendrology. An amateur can become very proficient in identification techniques of fossil-wood species. If cell-structure replacement is evident within a specimen, it can usually be identified by cutting and polishing an extremely thin section and examining it under a microscope. Like a fingerprint, the cell structure of a tree is its key identification element. Incidentally, well-agatized or -silicified thin sections are often quite translucent and lend themselves to slide mounting and subsequent projection with a 35mm slide projector. The accuracy of fossil cell replacement can be astounding—and very colorful. It should be noted that a fine collection of cut-and-polished slabs of fossil wood can be one of the most amazingly colorful and intriguing of paleomineralogical displays.

Locations of Petrified Wood

Insofar as locality types are concerned, there are perhaps four fundamentally different environmental situations that may eventually produce fossilized-wood specimens.

First and most common are sites where prostrate logs and branches apparently drifted or were washed into a water environment and sank to the bottom of a slough, lake, or shallow sea to become buried in sediments. The fabulous display of specimens in the Petrified Forest National Park, Arizona is a fossil deposit of this type.

Second are forests that were buried by a showering of volcanic ash or other debris and entombed in an upright position along with whatever trees may have been lying on the forest floor. The vast deposits in Yellowstone National Park, Wyoming and the Florissant Fossil Beds National Monument, Colorado are of this type.

Third are areas where viscous volcanic materials have apparently rained down on or engulfed logs or branches. Lava is a very peculiar material and can actually encase a tree and harden around its trunk before the wood is destroyed by fire. The result is a hollow mold, either of a prone or an upright tree. In some cases, the ancient molds later become filled with chalcedony or agate, and the result is a fossil log or limb cast. Casts will not show any annular rings or wood cell structure, but it is amazing how well the exterior appearance of the log or branch may be preserved. Casts may also occur in strata of nonvolcanic origin if the buried wood completely decays, leaving only a mold cavity. Several localities in Wyoming, Oregon, and Montana produce interesting casts.

The last category is an outgrowth of the preceding three. As millions of years pass, specimens may be washed into creeks and rivers by normal erosion processes and consequently propelled and tumbled for miles. They may become involved in some titanic flood and transported great distances or be gouged out

Figure 9-2. Chalcedony *casts* after small tree limbs or branches. Only the limblike shape and preserved markings on the exterior indicate that this is a fossil. These specimens are from Oregon, Montana, and Wyoming.

by glaciers and left hundreds of miles from their original site. Glaciers push huge masses of rock and debris in front of them, eventually leaving it as a moraine, marking the end of the retreating glacier's lobe. Sometimes unearthed specimens of fossil wood will reach an ancient floodplain or river delta as it enters a sea and once again become candidates for a second fossil environment! In any event, there are dozens of localities where specimens of wood are found as components in ancient gravels of one type or another. They are obviously *float*, showing signs of various degrees of natural erosion via tumbling for centuries in ancient stream or river channels.

Mechanics of Fossilization

We now come to the principles of *fossilization*. Scientists are not positive about the precise mechanics involved in turning a tree branch into a mineral form, yet the fundamentals are reasonably well established. Once the wood or any other living form becomes entombed under appropriate conditions, the stage is set for the possible replacement of the organic material by inorganic *permineralization.*

The first step is for the wood to become buried under conditions where air cannot foster decay, and presumably the material surrounding the wood prevents this or minimizes the potential. Eventually the sediments or volcanic ejecta become rock as more layers are deposited on top and the tremendous weight and pressure of the material compress silts into sandstones and shales, and volcanic ejecta into breccias or similar rocks. Many fossil-wood localities clearly indicate that perhaps thousands of feet of rock were once on top of the strata from which the wood is now eroding.

The second step results from slowly percolating mineral-laden waters coming in contact with the buried wood and, by a process not too well understood, gradually replacing the wood's cellulose structure with minerals. It is a strict replacement, or substitution, process, and the most common mineral involved is silica—quartz-variety chalcedony, agate, or jasper. The waters often include various minerals in addition to quartz such as manganese, copper, or iron, which, in turn, account for the astounding array of rainbowlike colors found in some fossil-wood specimens.

During the permineralization process, the cellular structure and the annular rings of the tree may be preserved in the utmost detail; however, wood can also petrify totally, leaving little or no indication of the former internal structure. The great proportion of fossil logs are simply *silicified*, which means that the replacing agent is a dull, brownish or grayish quartz, and the specimens have no use for lapidary purposes. However, silicified logs can preserve the most minute cellular detail. The forests of the Yellowstone area are of this type of mineralization. At other localities, the replacement mineral is again quartz but may be of the varieties known as chalcedony, agate, common or fire opal, and jasper. The intensely colored logs from the Petrified Forest are both agatized and jasperized, and cell structure is not commonly observed in this material.

The term *petrification* is often applied to wood, and quite appropriately. Fossilization, silicification, permineralization, petrifaction, and petrification all mean essentially the same thing. In the case of wood—and also other life forms under appropriate circumstances, such as dinosaur bones—the usual silica replacement is very hard. Most petrified wood is much harder than the surrounding host rock and consequently is easily exposed by normal erosion.

There are many odd aspects of fossil-wood deposits, as one might expect, with dozens of different localities available for study. Many areas have logs that are very poorly silicified, and the material involved erodes quickly upon exposure to air. Fossil wood clearly showing insect and worm holes and sometimes even the silicified insect is fairly common. Often the borings or tubes will be filled with clear, translucent chalcedony, while the wood is of another color. They make fascinating specimens. Fossil bark, knots, and tiny branches are very common at some localities, often accompanied by surrounding strata containing leaf imprints. Some fossil-wood localities also contain the fossils of shells, animal teeth and bones, corals, and even masses of compact swamplike reeds or tules.

Perhaps the prizes of fossil-wood enthusiasts are cycads and cones. Cycads were among the first flowering plants, contemporaries of the dinosaurs. They represent a stage of plant development that is between a tree fern and the palm family. Their bulbous, barrellike trunks have been found magnificently agatized at several localities in the United States—but they are uncommon. When they are cut and polished, they reveal a fascinating internal structure. Fine specimens have come from South Dakota, Wyoming, and Texas.

Pine cones are always a much sought curiosity item in fossil-wood localities and again are definitely not common. Fine little cones of ancient redwoods, firs, and other conifers, as well as various nuts and seeds, have been found in North Dakota, Wyoming, Idaho, and Oregon. However, the truly spectacular petrified pine cones are from the fossil forests of Patagonia, Argentina. Without a doubt, these rank among the world's most appealing, attractive, and fascinating fossil specimens. Anyone who sees one cannot help but be impressed

Figure 9-3. This is a fine example of fossilized worm-bored driftwood from Big Canyon, Newport Beach, California. Petrified wood is often found showing the tunnels of termites as well as the borings of sea worms such as the modern teredo.

by the amazingly accurate preservation of the cone—seeds and all. The cones represent several species of conifers and vary from 1" to 4" in length. They are brilliantly agatized and, when cut in half and polished, reveal the cone's internal structure in the utmost detail. I have inspected several of these cones in Buenos Aires—an unforgettable experience. This fossil-forest area is fairly remote in southern Argentina and is now a National Park. Collecting is forbidden.

It is also interesting to note that many other minerals or mineral forms may be involved in petrification besides the amorphous forms of quartz. A fossil log will often have cavities in it, either from rot and decay when it was entombed or from contraction of the wood structure once petrification began. These cavities—just like pockets in mines—serve as fine hosts for crystallized minerals. Clear-quartz crystals, smoky quartz, and amethyst have all been found in excellent museum-quality specimens *inside* fossil logs from various United States localities. As a matter of fact, one of the reasons that figured prominently in having the Petrified Forest set aside under the National Park Service was the uncontrolled dynamiting of huge logs in searching for amethystine-lined interiors! A mill had also been set up in the area to grind up the hard logs for sandpaper abrasives.

Some cavities in logs also contain well-crystallized pyrite, barite, dolomite, uranium minerals, and exquisite sprays of tiny stibnite crystals. Wood itself has been found replaced by lignite, limonite, calcite, and even native copper. Very unusual specimens have been found in Arizona that are thoroughly impregnated with native silver.

One of the most intriguing mysteries associated with fossilized wood is its apparent attraction for some of the uranium minerals. In the early days of uranium hunting in the western states, it was noted that fossilized logs in Utah were often good indicators of very rich deposits of various uranium minerals. Not only was the petrified wood often an "ore" by itself, but the area in the immediate proximity of the log had a uranium-mineral concentration as well.

Occurrence of Prone and Upright Forests

Now to some more specific details regarding two particularly notable displays of fossilized wood, one prone and one standing, both presenting entirely different geological stories.

One of America's most accessible major natural attractions is the famed Petrified Forest National Park located astride U.S. Route 66 some 20 miles east of Holbrook, Arizona. Millions of visitors have enjoyed this remarkable display of fossils as well as the ever-changing scenery of the Painted Desert.

The story of this deposit is well told in a series of fine dioramas and exhibits at the Park's visitor center and museum. The Park contains several forest areas—Blue Forest, Rainbow Forest, Jasper Forest, Crystal Forest, and Black Forest, located far out in the Painted Desert section of the Park. Actually, these are all part of the same geologic strata, and the "forests" are simply areas where erosion has exposed heavy concentrations of fossil logs.

The trees in the Petrified Forest are around 200 million years old and are found in the Chinle Formations of the Triassic Period. The strata which encase the logs are sedimentary and consist of a variety of very fine-grained siltstone, mudstone, and clays as well as coarser rocks such as sandstones and conglomerates. The beautiful array of subtle colors in the rocks of this area is essentially due to the presence of iron and manganese minerals. These have also produced the amazing array of rainbow hues in the agatized trees.

Figure 9-4. These are among the largest of the highly agatized and jasperized fossil trees in the Petrified Forest National Park, Arizona. Some 200 million years ago these abraded, prostrate logs washed into an ancient floodplain. In this swamplike area they sank into the muds and silts that eventually became their tombs.

Analysis of the geology of the area indicates that the now-fossilized logs were deposited in a swamplike environment on a vast floodplain. Ancient streams meandered across an essentially flat terrain, perhaps intersecting small lakes, lagoons, and ponds. Streams or sluggish rivers had their headwaters high in an ancient mountain range located in what is now New Mexico and southwestern Arizona. The streams that carried the logs into the forest area flowed northwest toward a huge inland sea that existed in the late Triassic and involved most of what is now southern California.

Although the area was a lowland, there must have been a considerable amount of water at times in order to wash in logs of the size now found fossilized. Almost all of the trees show positive signs of having been driftwood and are heavily abraded. Some of them washed in with portions of their root systems still visible. The forest is composed of thousands of *prone* logs, with only a few stumps found in place. Associated with the amazing log concentrations—

some of which might actually have been ancient logjams—are the fossilized remains of various freshwater reptiles, amphibians, fish, snails, and clams. These include oupelors, large salamanderlike creatures; phytosaurs, crocodile-like reptiles; various plant-eating fish; and ceratoduses, close relatives of the modern lungfish.

Fossil-wood specimens may vary from small branches (which are very uncommon) to trees that were probably over 200′ high and about 10′ in diameter. It is estimated that approximately 90 percent of the fossilized trees in this deposit are of a group called Araucarioxylon. The closest living relative is the araucaria, or monkey-puzzle tree, now only found naturally in various sections of South America. Another fairly close relative is the Norfolk pine, and both of these trees are used for decorative planting throughout the United States. Two other genera are found in the forest, Woodworthia and Schilderia, as well as indications of a wide variety of ferns and various species of cycad-like plants called bennettites.

A number of complicated geologic cycles have affected the area since the original entombment of the logs. A huge lake, called Lake Bidahochi by geologists, formed in the region due to a general uplifting of northern Arizona. This drastically affected the drainage patterns. of existing rivers. Eventually this prehistoric lake was drained and played an important part in developing the Colorado River and the Grand Canyon as we know them today.

Only a few million years ago volcanos became active in the buried forest region and added volcanic ash and lava to the ever-mounting deposit of sediments accumulating atop the buried logs. Eventually over 3,000′ of rock were deposited. Then erosion began its relentless stripping away of the land and gradually cut deeply into the Chinle Formations, exposing the new fossilized trees. Erosion continues today, and the National Park Service estimates that 1/4″ to 1/16″ of soil is removed from the area each year, depending on the degree of slope involved in each area. This process continually uncovers additional fossils, and there can be no doubt that hundreds of logs await exposure as they lie buried in the mesas and hills of the Park.

The so-called pedestal logs of the Park are fascinating examples of erosion at work. The principle is simple—the logs are extremely hard; the surrounding rock is not. The soft rock gradually erodes away on all sides of a log and often leaves it perched atop a narrow ridge of unexposed rock. In essence, the log provides a protective capping for the softer rock and produces a pedestallike formation. Eventually erosion wins and the log will topple, perhaps rolling down a hillside once again to begin the same process.

Now let's discuss another major fossil forest that is entirely different in terms of burial and that also reveals a multiple story. There is probably no place on this planet wherein more unusual natural phenomena and beautiful scenery are combined than in Yellowstone National Park, Wyoming. However, few people realize that Yellowstone contains the world's largest deposit of fossil trees! If there were no other phenomena in the Yellowstone region, part of the area would undoubtedly have come under national-park or -monument status by virtue of the incredible fossil forests.

The fossil forests of the Yellowstone are distinctly different from Arizona's. Here the majority of trees stand upright, just as they grew before being buried by successive volcanic eruptions. Not only are most of the trees upright, like the columns of an ancient temple, but in several areas there are *series* of forests, buried on top of one another.

With but one exception the fossil forests of the Yellowstone are reasonably inaccessible, with most of the truly spectacular areas quite remote or reached only by difficult climbs. Fortunately, this is for the best when you consider what has happened to the only readily accessible area, the well-known Petrified Tree near Tower Falls. On a grassy hillside not far from Camp Roosevelt stands a lone 12′ petrified tree, completely surrounded by a high iron fence. This is all that remains of a small forest of some thirteen standing trees or stumps and is a monument to vandalism. Souvenir hunters hacked and pounded the rest of the fossils into oblivion.

Another factor of great interest regarding the Yellowstone deposits rests with the astonishing lifelike appearance of the fossil trees, stumps, bark, root systems, and fragments. Sometimes it is actually difficult to tell at a distance if a tree is fossilized or merely a dead snag. Prostrate trees—and there are many—often weather in a manner that virtually duplicates kindling wood.

Figure 9-5. There are several of these pedestal logs within the Petrified Forest National Park, Arizona. They are the product of differential erosion—the log is petrified and very hard; the surrounding rock is not. The log acts as a protective capping for the rock immediately under it. As the surrounding rock is eroded away, the log remains perched atop its protected pedestal. Eventually erosion will win and the log will topple.

The extensive fossil forests of the Yellowstone region (both in and out of the Park boundaries) appear to be contiguous. The composition of strata entombing the fossils is virtually identical in the various forest areas; however, the degree and type of mineralization differs from locality to locality, particularly north of the Park proper.

The fossil forests were buried during the Miocene epoch of the Cenozoic Era, which gives them an age of perhaps 50 million years. They were buried by a *series* of eruptions from volcanic vents, presumably located to the northeast of the present Park boundaries. As the climate changed during the Miocene, a variety of trees lived in this region, but in all probability not at the same time. The forests contain redwood, pine, oak, buckthorn, laurel, bay, sycamore, and even magnolia. Of the fossilized trees, however, some 95 percent belong to the conifers.

Figure 9-6. Yellowstone National Park in Wyoming, Idaho, and Montana contains vast deposits of standing fossil forests. Buried some 50 million years ago by a series of volcanic eruptions, over twenty-one separate forests stand on top of one another. These are typical fossil pine tree trunks on the slopes of Specimen Ridge.

Figure 9-7. The Grandfather, a huge fossilized stump of an ancient redwood. This specimen is located in the lofty and beautiful Tom Miner Basin, north of Yellowstone National Park. There are many fossil-forest areas outside the Park boundaries.

A locality such as that of Specimen Ridge is like reading an open textbook on geology—with each successive stratum a page in the fossil story. An eruption of the volcanic vent or vents hurled enormous quantities of mud, ash, and sharp, angular rock fragments high in the air to rain down upon the landscape and its currently prevailing forest. The forests were not buried by lava flows, nor do they show evidence of being blasted to the ground by violent explosions—as a matter of fact, just the opposite. As the heavier fragments fell, they might have snapped off branches, or perhaps the sheer weight of finer debris accumulating on a limb would break it off. So remarkable is the state of preservation of many standing trees that you can actually see where branches were broken off, tearing downward into the bark. This created a scar exactly like that caused by accumulated snow on a living tree.

As an eruption progressed, finally only the bare, stripped, standing trunks were left to be engulfed by the showering ejecta along with the litter of branches and leaves on the ground. When the volcanic rumbling ceased and the sky cleared, instead of a green, lush forest there was an arid, steaming, barren plain—stirred only by hot winds drifting the newly fallen ash. Perhaps a snag broke the desolate surface, the only reminder of the forest buried beneath.

During the ensuing years the rains and snows saturated the new ground, and eventually a new topsoil began to form. Winds carried seeds into the devastated areas, and soon a new flora took root to flourish until a recurrence of the volcanic activity buried it in turn. Meanwhile, the pressure of the huge weight of material was consolidating the ash, mud, and rock fragments into hard breccia. In their rock tombs the trees underwent the slow process of petrifaction, and each minute cell of their structure was replaced by hard, gray or brown silica carried in solution by percolating waters. Trees in these deposits are not highly agatized like the colorful trees in Arizona.

How many times this cycle of growth and burial was completed no one knows, nor do we know how many generations of trees existed on any one given landscape before it was buried by a new eruption. Erosion, however, has opened up many buried secrets. In many sections of the Yellowstone region deep canyons and valleys have been cut into the extensive tree-bearing strata. As the softer breccia erodes away, it leaves the hard fossil-tree forms behind. Consequently, the trunks are left standing erect upon lofty ledges, in sheer cliffs, on hillsides, or atop ridges. There are several localities where ten different levels of buried forests may be seen, and some exposures in the Specimen Ridge formation yield eighteen to twenty-one levels of forests buried on top of one another! Leaf and frond fossils representing over one hundred fifty species are abundant in the fine ash layers in certain localities, but to my knowledge no vertebrate remains have yet been found in positive association with the fossil trees. It is interesting to note that intermixed in these multilayered fossil-tree-bearing strata there are layers containing no fossil remains. We have no idea what type of landscapes they may represent.

The fossil forests within Yellowstone Park are extensive, and the entire northeastern and northwestern sections of the Park contain numerous localities. There are also many extensive localities outside the Park boundaries.

Chapter 10

CAVES AND CAVERN FORMATIONS

Caves have always fascinated man, seemingly offering him an irresistible and eerie appeal steeped in mystery. Prehistoric man used these formations for his first shelters, and some of the world's truly primitive tribes still live in caves. Pirates, robbers, escaped convicts, mountaineers, and refugees have rendezvoused, plotted, or hidden in caves. Treasures, from gold coins to sacred effigies, have been secreted in caves. Tom Sawyer and Becky Thatcher were lost in a cave, as many amateur and professional spelunkers have been. Ancient peoples buried their dead in the silent, inky depths of caves, perhaps the first mausoleums, and today modern man maps the extent of known caverns for bomb shelters or possible sites for the disposal or storage of atomic wastes. No matter how you look at it, caves have always been an integral part of man's experience, ancient and modern, and one of geology's oddest phenomena.

Speleology is the name given to the scientific study of caves, and people who pursue this science or adventuresome hobby as professionals or amateurs are called *spelunkers. Speleothem* is a general term that designates any type of decorative cavern formation.

There are many practical rules of the game for cave exploring, or spelunking. Never explore a cave unless you are accompanied by someone reasonably familiar with basic cavern-safety procedures. For example, caverns should not be explored below surface level if it is raining or if rain is threatened. Some caverns become flooded during storms, and you might be trapped. Similarly, team exploration is positively recommended: getting lost in a cave all by yourself is the epitome of a lonely experience. Many of America's most beautiful caverns have been discovered and mapped by amateur speleologists. The thrill of finding a room that no human has ever seen before is hard to express.

Types of Caves

Geologically speaking, there are different types of caves. Perhaps the most common is a simple overhanging ledge of rock that is the product of surface erosion by wind and water. Today such ledges shelter birds, wild animals, and humans who get caught in hail, rain, and snowstorms. Perhaps these are more accurately *rock shelters* than caves per se.

Caves also occur frequently in lava beds and are called *lava caves*, or *lava tubes*. They may vary in size from a sewer pipe to a subway! These tubes form in pahoehoe-type lavas which flow like thick, frothy molasses from cracklike vents in the earth's crust. As they flow, the lavas cool and harden first on the surface and later on the sides of the flow, and when the plastic lava inside the flow continues to drain out, a conduit, or tube, is left behind. As the tube is created, the cavity fills with superhot gases from the molten floor, which remelt and glaze the roof and sides of the tube with astonishing lava pendants, walls that appear glazed and dripping, and odd stalactitelike formations called *lavacicles*.

There are several areas that contain lava tubes. Many miles of lava caves occur within Lava Beds National Monument in northeastern California. Catacombs Cave may be explored with perfect safety for about 1½ miles, and Valentine, Indian Wells, Sunshine, and Sentinel Caves are equally fascinating. The monument also includes Post Office and Dragon's Head Caves, both of which are huge lava tubes. Nearby is a series of lava caves stacked one atop the other called Bear Paw Caves. The lower of these caves (which is quite a climb) is called Merrill Ice Cave and features a permanently frozen ice river and a beautiful ice waterfall. Tubes or even limestone caverns that exist in areas where win-

Figure 10-1. This is a typical scene in Lehman Caves National Monument, Nevada. Located on the eastern slopes of glacial-carved Wheeler Peak, Lehman Caves is outstanding for the profusion and variety of its cavern formations. This scene shows *stalagmites*, *stalactites*, *flowstones*, *draperies*, *bacon rind*, and *columns*.

ters are severe may serve as natural deep freezes and contain beautiful speleothem formations of glistening ice caused by the dripping of water into the cave and its subsequent freezing. Perhaps the most dramatic cave of this type is the Crystal Ice Cave in Idaho, some 150′ below the surface, in which huge, transparent icicle columns may reach 25′ in length. Visitors to Hawaii Volcanos National Park are always entranced by the very large Thurston Lava Tube. Small tubes are present in Craters of the Moon National Monument, Idaho, and there are many others in Oregon, Washington, New Mexico, and Alaska.

As indicated above, an ice cave can be a lava tube or even a limestone cavern if it supports icicles and other speleothem-related formations of ice. However, a real *ice cave* is found under a glacier. Glaciers are simply slow-moving rivers of ice, and water flows underneath all of them. The water, coupled with air currents, fashions beautiful ablated ice caves, which extend underneath a glacier for various distances from its snout. The Paradise Ice Caves under the Paradise Glacier of Mt. Rainier National Park are the most accessible. The depths of a glacier ice cave are a chilling and eerie experience in blue!

Related to ice caves but with a different cause are the steam-filled caverns that exist atop Mt. Rainier within its perennial glacial ice cap. These are caused by escaping hot geothermal vapors. The openings are large—one even includes a small lake—and have been well explored by geologists. Other caverns are suspected to exist in the icy caps of Mt. Baker, Mt. Adams, and other volcanic peaks of the Northwest, but the presence of noxious hydrogen sulfide and other gas fumes has prevented exploration.

Caves are sometimes found in ancient hot-spring areas and are called *travertine caves*. Various localities in Colorado, Arizona, New Mexico, and Wyoming have caves of this type, wherein you can explore the plumbing system of a dormant hot spring. However, caves located in still active thermal areas can be dangerous due to the accumulation of gases. A rule of thumb: if a cave smells bad, don't go into it.

Other caves are initiated by the weakening action of a spring in a cliff wall, which combines with gravity and external erosion—wind, rain, frost—to gradually spall the rock away around the seepage. Most of the huge caves in the sandstone cliffs of northern Arizona are of this type, and many, such as Keet Seel and Betatakin, were used as ancient village sites by the prehistoric Anasazi Indians.

There are a few so-called caves that are the product of circumstance, weathering, and rock rubble! *Talus caves*, as they might be called, are simply the winding and tortuous passageways between a jumble of huge boulders in a rockslide or talus slope.

Littoral caves are frequently found along seacoasts where the relentless carving action of waves between low- and high-water levels has produced caves in cliffs or promontories. Sea Lion Caves, Oregon and the caves at La Jolla, California are excellent examples. Some littoral caves are now found far inland as reminders of ancient seas that existed eons ago. In a similar manner, river waters may undercut and gouge an amphitheaterlike opening in a mass of resistant rock at a sharp bend in the river. Eventually the river will win and the rock will be penetrated or eroded completely away.

Perhaps the most accurate use of the term *sea cave* refers to a cavern opening and subsequent chambers under the sea, accessible with scuba gear; many limestone-cavern openings occur underwater in various areas. The sea can also create underwater caverns by normal erosion processes in fissures and cracks in other types of submerged coastal rocks.

Spring caves are huge, flowing springs found in Florida, on islands in the Caribbean, and in Yucatan, Mexico. These large, clear, cold springs issue from a limestone bedrock and are actually the openings of water-filled cavern systems, or, if you prefer, the outlets for underground rivers. Many of these have been explored by divers equipped with appropriate scuba equipment. At least one such spring cave, the Wakulla Springs Cave near Tallahassee, Florida, has furnished the bones of Pleistocene mastodons and tapirs buried in the silts on the spring floor.

The most fascinating of all caves is the *limestone cave*, or more correctly, the type of cave that occurs in limestone or marble strata formations. These caves are not particularly rare, since they may occur wherever limestone deposits are found throughout the world; their extent, size, and beauty differ considerably. They may vary from a few dozen yards in length to an interlinked system of levels over 100 miles long. There are hundreds of limestone caves in the United States, which has developed more caves for tourists than any other country. Other nations in which there are notable caves are Mexico, Australia, Switzerland, Yugoslavia, Lebanon, Brazil, France, Czechoslovakia, China, and the U.S.S.R.

A cavern may be either a simple series of interconnected rooms and passageways at approximately the same level or a vast system of interconnected corridors and rooms at different levels. For example, at Carlsbad Caverns National Park there are at least three major levels of interconnected caverns, the lowest of which is 1,320′ below the surface. The beautiful Luray Caverns of Virginia have four separate levels of subterranean chambers. Another type of cavern, typical of many in the French Alps, is called an *alpine pit*. These are found in radically tilted limestone strata, and, as the name implies, the cave is essentially a deep, vertical shaft penetrating the rock. Caves of this type are found in California's Sierra foothills and in Montana.

It is interesting to note that in some caves that are a series of interconnected levels, different phases of cavern formation may be in progress at the same time. The lowest levels may have free-flowing water in them and are in the process of being excavated, while the upper levels are free of such waters and are now being decorated by the dripping and evaporation of mineral-laden waters. While the basement is being dug, the top floors are being decorated!

Cave Development

In essence, the formation of a limestone cave is not particularly complicated, but it may take millions of years to first produce the basic cavity and then to decorate it with cavern formations. Caves can be said to have three growth phases: excavation, decoration, and dormancy.

Beds or strata of limestone or marble (marble is metamorphosed limestone) are commonly faulted, cracked, and fractured by earth movements, and certain members of the strata may be much softer than others. It is along these planes of weakness that caverns develop. Water is the excavator.

Rain, snowmelt, and other groundwaters containing carbon dioxide seep or flow downward along cracks, joints, fault planes, and fissures. This water actually constitutes a weak form of carbonic acid and slowly dissolves the limestone. Gradually this dissolving action enlarges the cracks, joints, and crevices into passageways, rooms, or huge halls. These barren, dissolved areas are often called *solution channels.*

The rate of erosion for a cave is relative, depending upon the amount of water involved, the length of time, the rate of flow, and the resistance of the particular limestone. Some caves have apparently been created by an underground stream or river flowing along the line of least resistance deep in the earth. Several caves still have active streams or rivers flowing in them. Other limestone caverns may have been formed at the saturation level for groundwater. After solution channels have been excavated, rooms may fill up with very slow-moving water. This water acts on the walls of the room, gradually expanding the chamber by dissolving the walls and causing the ceiling to collapse. The ceiling rubble is then attacked by the water, which in turn dissolves this material and carries it away. This is somewhat of an oversimplification, but gigantic rooms, such as those in Carlsbad Caverns, have been created by this process.

As might be imagined, due to the variables mentioned, caves are found in various stages of excavation, varying from the tiny circuitous galleries of Jewel Cave in South Dakota to the incredible and immense halls of Carlsbad Caverns.

Eventually the prime dissolving waters leave the caverns. This may be due to geological uplift of the region, causing the water to drain away, or perhaps the original source is diminished, terminated, or diverted. Then air fills the chambers—most accessible caverns have an abundance of fresh air in them. Air and water are essential to the second phase of cavern development, decoration.

In one sense, nature now attempts to refill the opening created by the excavation phase. Water continues to seep into the cave via cracks, joints, bedding planes, and faults. As these groundwaters percolate downward through the rock strata above the cavern opening, they continue to dissolve the limestone and carry with them calcium carbonate in solution. Upon entering the cavern at the roof, wall, or even floor level, the water comes into contact with the air and evaporation takes place. Consequently, wherever cavern walls or ceilings leak, mineral deposition takes place by evaporation. Thus begins the decorative process that produces the weird and beautiful formations that festoon many caverns and transform barren rooms into fairylands.

Cavern Formations

Caverns are found in varying degrees of decorative development, and again, this depends on many variables—the rate of water flow, the rate of evaporation, and the amount of calcium carbonate held in solution. But above all, it is the seemingly eternal dripping of water in total darkness that creates cavern beauty. The following are a few of the basic types of cavern formations.

Where water enters a cavern through the ceiling, the processes of bathing, evaporation, and deposition produce a *stalactite.* Stalactites always "stick tight" to the ceiling. Frequently a stalactite is supplied with a surplus of water. In other words, all the water cannot be evaporated at the ceiling level, and consequently some drips off the growing stalactite to the floor directly below. This begins a formation that grows upward from the floor toward the ceiling, a *stalagmite.* A stalactite is usually slender and graceful; a stalagmite, stocky and thumblike. Both can grow to immense size. Actually stalactite and stalagmite are growing toward each other, with the stalactite furnishing the water for the stalagmite's growth. Often these two features meet and form a *column.*

As would be expected, stalactites, stalagmites, and columns come in all sizes and shapes from tiny, hollow, crystalline, strawlike tubes to 40′ giants such as in Carlsbad's Big Room. Incidentally, a huge stalagmite may not necessarily have a large stalactite above it. The ceiling opening may have such a large water supply that very little mineral is deposited at the ceiling level, and most evaporation and deposition takes place at the floor level. The reverse of this can also be true, with a stalactite achieving a maximum balance between water source and evaporation and growing almost to the floor. Such features may often have lovely *terraced pools* beneath, with scalloped and fluted edges—all the results of evaporation of the mineral-laden waters on the floor.

There are many variations on the stalactite-stalagmite theme, which have self-explanatory names. Caves commonly have cracks in the ceilings that may be active for their entire length. Dripping waters from such an extended source produce beautiful ribbonlike or streamerlike stalactite formations called *draperies*, or *curtains*. Those that are exceptionally thin and translucent may be called *bacon rind*.

Active seepages in the wall of a room may build up great mounds of cascading deposit. Such formations can be extremely large and are usually referred to as *flowstones*. The so-called frozen waterfalls are usually formations of this type.

Perhaps the oddest and most interesting of all cavern formations are called *helictites*. These are curious twisted and branched formations that occur on the side walls of caverns or even on the sides of stalactites. The mechanics of helictite formation are not clearly understood. The delicate branching is perhaps a combination of released carbon dioxide in the water and shifting air currents. The sheer beauty of helictite formations is difficult to describe, particularly if the waters contained mineral impurities and tinged them faintly in blue, green, yellow, or red hues.

Some wall surfaces in the Queen's Chamber of Carlsbad are so heavily encrusted with helictite formations that they resemble thorny thickets. Similarly, some of the world's most beautiful mineral specimens are the fabulous helictite-encrusted stalactites on display in Timpanogas Cave.

It is not uncommon for *all* types of cavern formations to show intermittent or interrupted growth, often yielding a gnarled appearance, or what might look like an erratic alignment. This is caused by a fluctuation of the water supply, as well as differentials in mineral content and perhaps even the effect of shifting wind currents. This is quite evident when a stalactite or stalagmite is cut and polished to reveal the structure of concentric rings—much like tree rings. The thickness and even the color of the rings differ considerably. However, the rings—unlike the growth rings in a tree—are unequivocally *not* added on an annual basis. A ring may represent a few months—or several hundred years. There is no standard rate of growth for cavern formations. All cavern sites differ in water supply as well as in the amount of calcium carbonate in solution: each individual formation has its own supply. A very small formation might, in fact, be thousands of years older than another huge one nearby.

Another unusual and uncommon cavern formation is called a *shield*, or *palette*. These are round, disklike formations protruding from cavern walls. Beneath the disk there is usually a shroud of slender stalactites, which often makes the formation look like a parachute. Particularly beautiful examples occur in Lehman Caves, Nevada and in Grand Cavern, Virginia. The mechanics of shield growth are not fully understood, and they remain an interesting and perplexing puzzle.

Among the most unique of all cavern formations are *pisolites*, *oolites*, or *cave pearls*, as they are commonly called. Under certain circumstances, water dripping into shallow pools will begin to deposit calcium carbonate in concentric layers around grains of sand, fragments of insects, or bat bones. The constant dripping spins the formations, and loose marblelike spheres start to form! Most cave pearls are small, perhaps ½" in diameter, but occasionally golf-ball-size specimens are found. Cave pearls frequently occur in "nests" of a dozen or more units and are probably the most fascinating of all speleothems. Interestingly enough, very similar formations occur in the basins of

Figure 10-2. A typical frozen waterfall of calcium-carbonate dripstone, or flowstone, adorning one of the walls in the Hanson Cave section of Timpanogos Cave National Monument, Utah. Mineral-laden waters dripping on the slope of the wall evaporated and deposited these draperylike stalagmite speleothems.

certain small fountain geysers in Yellowstone National Park. In this case the geyser "eggs" are composed of silica, or geyserite, deposited by the splashing action of falling hot water.

Due to the many variables (amount of water, air currents, mineral content in the water) involved, cavern formations can occur in a bewildering variety of fantastic shapes and sizes. They may be graceful, gnarled, long, thin, thick, cup-shaped, beet-shaped, nodular, grapelike, coralloidal, sponge-shaped, fluted, terraced, stepped, cupola-shaped, streamerlike, corrugated, cone-shaped, canopy-shaped, or lily-pad-like.

The most common mineral deposited in caverns during the decorative phase is calcium carbonate—calcite—and sometimes its chemical twin, aragonite. Eons of evaporation leave tiny layers of this material behind, gradually fashioning the various cave formations.

Figure 10-3. Sparkling *helictites* of pale pink and white translucent aragonite adorn the sides of stalactites and the walls of the Chimes Chamber section of Timpanogos Cave National Monument, Utah. This series of small, interconnected caverns high in the Wasatch Mountains contains one of the world's most remarkable displays of helictite formations.

Unfortunately, the fanciful descriptive terminology of cavern mineralization in common use is quite confusing. Calcite occurs as a secondary deposit in caverns under various conditions that have fostered terms such as dripstone, rimstone, caverock, grottorock, shelfstone, cavestone, flowstone, cavernstone, and more. All are calcite.

When the formations are very hard and compact and show distinct rings or layers, as is usually the case in stalactites and stalagmites or in flowstone, the material is sometimes called "onyx." This term is a takeoff on the true use of the word, which refers to a straight-banded type of chalcedony. It might be more accurate to call the material "cave onyx" or, as the well-known material from Mexico is called, "Mexican onyx."

Figure 10-4. The Parachute, a *shield*, or *palette*, speleothem in beautiful Lehman Caves National Monument, Nevada. This cave displays several of these very uncommon and puzzling shroudlike formations. They are a cavern oddity that has not been thoroughly explained.

When the deposits are very large, the term "travertine" is sometimes used. Here again, a certain amount of confusion is involved. Travertine is usually associated with the deposits of ancient or modern hot springs. Examples are the great commercial travertine deposits near Rome, Italy and Gardner, Montana. Such material is normally somewhat porous and is frequently used for interior building walls and other decorative purposes. Nevertheless, some descriptive literature on caves refers to formations of travertine, which may be chemically correct but depositionally open to question.

In some caves, particularly in tiny Timpanogas Cave, Utah, aragonite has crystallized in magnificent acicular bursts encrusting helictites or stalactites of the same mineral. The brilliance, incredibly delicate structure, and subtle coloring of the Timpanogas aragonites is quite unique.

Figure 10-5. The Monarch of New Cave in Carlsbad Caverns National Park is certainly one of the world's largest stalagmites—if not *the* largest. It is entirely freestanding. This behemoth is over 15′ in diameter and reaches some 60′ in height. The tall, thin stalagmites in the foreground are called *totem poles*, and the slightest tap will cause them to vibrate. This undeveloped cave is not open to the public.

A truly odd formation seemingly originates from small cracks in the hard rock wall of a cavern, and the subsequent formations, called *oulopholites*, look like they have been squeezed out of the crack! Gypsum is the most common mineral seen in these odd extruded forms and is given descriptive names such as "ram's-horn gypsum," "gypsum flowers," or "gypsum lips." Gypsum may also occur as massive platelike deposits on a cave floor; as huge, well-developed, and spectacular swordlike crystals (as in the Cave of the Swords, Naica, Mexico); or as amazingly delicate clusters and sprays of needlelike filamentous crystals. This material is also called "gypsum hair" and "cave cotton." The various gypsum formations mentioned are frequently encountered in mining activities and have all been formed while caverns or cavities were filled with saturated solutions containing gypsum, which subsequently crystallized.

Figure 10-6. The Tear is a spectacular and unusual stalactite formation in New Cave (also known as Slaughter Cave) in Carlsbad Caverns National Park, New Mexico. This speleothem is very large and is composed of glistening white calcite. Eventually it will grow to be a column, such as the formation on the right. There are many caves within this National Park in addition to the huge Carlsbad Caverns proper.

Calcite *crystals* are not uncommon in caves, particularly in Jewel Cave National Monument in South Dakota and in many caves in Virginia and Missouri. Such crystals were deposited when the rooms were full of solutions containing calcium carbonate. Scalenohedrons (dogtooth-type) and nailhead-type crystals are the common forms. The basic color of these cavern formations is either white or brownish. Additional coloring in cavern formations (exclusive of lighting) is due to various impurities, such as mud and other minerals, that may be contained in the evaporating waters. Tints of yellow, orange, dark brown, amber, red, and blue-green are not uncommon.

Incidentally, visitors are often mystified by the green coloration on wet formations that may be near incandescent lights in caves. The green is algae, the spores of which have been carried into the cave by air currents. The spores have found a spot where they can thrive, nurtured by moisture and the warm radiation from the light bulb.

From the standpoint of the mineral collector, it should be noted that certain terms used in speleology are also shared with descriptive mineralogy. For example, stalactitic, coralloidal, mammillary, spherical, botryoidal, pistolitic, crystalline, and capillary are used at times to describe certain speleothems and are also commonly used to describe the physical forms in which many minerals occur.

Openings, vugs, and cavities in ore bodies (miniature caverns, if you prefer) offer idealized environments for mineral deposition, either as crystals or as any of the numerous other physical forms in which a mineral may occur. Stalactitic specimens of malachite, marcasite, smithsonite, celestite, rhodochrosite, chalcedony, pyrite, limonite, and prehnite are not uncommon. In the early 1900s the Copper Queen mine at Bisbee, Arizona produced some stalactites of deep emerald-green malachite liberally sprinkled with brilliant azure crystals of azurite. The specimens are positively spectacular and quite valuable.

Coralloidal simply means "corallike," and dozens of minerals may occur in this form. The same holds for *mammillary*, which refers to contiguous rounded surfaces, and its companion term *botryoidal*, which means grapelike. Many handsome specimens are found in these forms, particularly of hematite, malachite, azurite, mimetite, smithsonite, and chalcedony.

Under no circumstance is the data presented in this chapter designed to encourage the collecting of specimens in caves, grottoes, and caverns. Quite the opposite. Cavern specimens belong in their original environment, the only place where they are meaningful. Specimens can be acquired from commercial dealers, who hopefully have obtained them from quarry or mining operations, which infrequently encounter small caverns and would probably otherwise destroy the speleothems.

Caves and caverns are definitely not good sources for mineral specimens, regardless of what you may have been told. Cavern specimens are fascinating when viewed in the cave; once removed, they become essentially meaningless, seldom prove attractive, and often dry out and alter into completely useless material.

All of the caverns discussed are under the protection of the National Park Service or some branch of state government or are privately owned and developed. Vandalism in unprotected and undeveloped caverns should not be tolerated, and it occurs in two forms: outright disfigurement and destruction of a cave and the collecting and removal of speleothem formations. I have visited unprotected caverns that have literally been stripped of every vestige of cavern formation. Aesthetically and scientifically they have been looted and destroyed and now are as interesting as an empty warehouse. What had formally been a fantasyland of irreplaceable formations has been transformed into a shambles.

Today no one questions the fact that conserving the odd, unique, beautiful, and spectacular aspects of our natural environment is essential to the heritage of the generations to come. Our country is blessed with a phenomenal variety of spectacular scenery. We can readily see and are anxious to preserve features such as petrified forests, natural bridges and arches, prehistoric cliff pueblos, and dinosaur remains. The aspects of our natural heritage that lie *below ground* deserve equal attention. There is outdoor wilderness and *underground* wilderness, both needing rational conservation and protection.

Since the early 1940s, the National Speleological Society, Inc., an exceptionally well-organized group of both professional and amateur speleologists, has been involved in a detailed study, mapping, and exploration program of the nation's cavern resources. One particularly meaningful section of their policy states that ". . . caves have unique scientific, recreational, and scenic values; these values are endangered by both carelessness and intentional vandalism; these values, once gone, cannot be recovered; and the responsibility for protecting caves must be assumed by those who study and enjoy them." These objectives deserve the wholehearted support of anyone interested in minerals, fossils, and the lapidary arts.

There is also a very good case for discouraging not only the collecting but also the purchase of cave minerals. The point involved is that selling speleothems instantly places a market value on them and consequently may encourage others to ruin a cavern purely for a few dollars! Assuredly, a polished slab of flowstone can hardly be equated with a broken gash in a beautiful cavern environment.

There is also the fact that hundreds of undeveloped caves are on *private property*, and the removal of speleothems can be considered as malicious vandalism or theft. There have been several occurrences of outraged property owners who have closed the caves to all. Unfortunately, this is not a new story to those who have seen many fine mineral, lapidary, and fossil localities closed due to the thoughtlessness of a few.

A cave per se is a feature to be enjoyed but also to be studied scientifically. The entire cave environment—the rooms themselves, the speleothems, and the various life species often found in caves—should be left untouched. In one sense, the cave environment is a fragile, delicate thing, and scientists still have much to learn about cave mechanics and cave life.

Chapter 11

MINERAL SPECIMENS AND VALUES—WHAT'S IT WORTH?

There are hundreds of thousands of specimens of the different minerals in collections, both private and museum, all over the world. How are their values established? What are the differentials? This dilemma was perhaps best summarized by two recent conversations, one with a more or less folksy but well-experienced mineral dealer in southern Illinois and another with a very sophisticated dealer in fine crystallized minerals in southern California.

"How do you estimate the value of a specimen?" Pause. "Well, it ain't easy. Too many changing things you gotta think about."

Same inquiry. Pause. "Well, there are a great number of variables to take into consideration, and it depends a lot on both the dealer and the collector's criteria of quality."

Both gentlemen, have in essence, said the same thing. Values are the product of a fascinating and perhaps odd composite of many factors.

An interesting aspect of value in general can be found by reading *The World's Finest Minerals and Crystals* by Dr. Peter Bancroft. In this interesting work the author and a panel of judges and consultants selected seventy-six examples of remarkable mineral specimens from worldwide collections in both private and museum hands. Many readers might well contest some of the selections, yet no one can deny that the specimens included (accurately illustrated in color photographs) are truly superlative or exceptionally rare and unusual.

The criteria used for evaluating the specimens in this unusual book are very interesting. They were as follows: crystal perfection, preference of crystals on matrix specimens, general aesthetic appeal of the specimen, association or combination of minerals in an unusual or attractive manner, unusual color quality, brilliant luster, and size—with an obvious preference for smaller specimens of great perfection.

All of these factors have been covered in reasonable depth in the foregoing chapters of this book. They help to indicate that determining the ultimate value of any mineral specimen, superb or poor, is always a composite of experience, knowledge, and skill. A good memory also helps!

Although the primary reason for collecting specimens is essentially pleasure and enjoyment, the fact remains that as a collection grows, it is definitely worth something. Due to the degree of enthusiasm, personal preference, depth of specimen types, and basic specimen quality, collections may vary from the casual to the professional and be worth a little or a great deal.

Strangely, and somewhat contrary to what many collectors might believe, many of the odd and unusual specimens covered in this book have very little monetary value. There are some exceptions, of course, since many of the best pseudomorphic specimens (depending upon the particular minerals involved) may command high prices, as would quartz crystals with spectacular multiple inclusions.

However, cone-in-cone specimens, concretions, speleothems, fulgerites, sand spikes, and many types of geodes have value only as geologic curiosities. As such, they command modest prices and are perhaps encountered most frequently in teaching situations. Nevertheless, many mineral or fossil collections will definitely include examples of various specimens described in this book, and many collectors have a special suite of odd and unusual specimens.

In regard to materials within the lapidary field, such as agate, jasper, opal, jade, turquoise, and petrified wood, it is virtually impossible to document any particularly valid price structure. Lapidary materials are generally not expensive (as compared with superfine crystallized minerals), yet exceptional specimens of jade, plume agate, or multicolored wood may bring prices in the dollars-per-square-inch bracket or, as in the case of opal and turquoise, the dollars-per-ounce category! Precious and semiprecious gemstones almost automatically have a well-established market price determined by wholesale and retail outlets. Consulting the various magazines that emphasize lapidary materials will quickly reveal the current price structures of hundreds of different items, ranging through synthetic star sapphires, plume and moss agates, scenic jaspers, opals, thundereggs, fire agates, petrified wood, tanzanite, diamonds, turquoise, and jade.

It is important to keep in mind that there are many levels of, degrees of sophistication in, and even reasons for collecting minerals, fossils, or lapidary materials. In essence, there are two fundamental types of collectors: those who collect or work in the field for the sheer enjoyment afforded by a hobby embodying both relaxation and learning activities and those who might best be referred to as trophy hunters. Exhibiting specimens at shows is an exceptionally worthwhile activity, but in recent years some aspects of "competition" have been carried to the point of absurdity. Judging shows has become a quagmire of involved rules, regulations, and split decisions, with preference frequently going to the most well-heeled collector. The trophy-hunter syndrome has produced a small coterie of international collectors seemingly obsessed by competition. Enjoy these fine collections when they are on exhibit, but don't become disenchanted if your own fine collection never takes best in show or a first. Collect to learn, share, and enjoy, not to win ribbons, or you may find yourself in the psychiatrist's office. A collector really does not need to become a frustrated neurotic, obsessed with absolute specimen perfection and rarity, in order to enjoy mineral collecting. There are always a few collectors who seemingly and perhaps justifiably can survive only on the superlative. However, between this level of collecting and the proverbial helter-skelter rock collection there is a great deal of latitude. Perhaps this can be called the enjoyment zone.

A word about so-called perfect specimens. Absolute textbook-illustration perfection in a mineral specimen is not particularly common. However, it is astonishing how many truly fine, unblemished specimens may either be found or purchased if you become familiar with the variables and techniques involved. To repeat a case in point, in any selection of fifteen wulfenite specimens offered for sale at $10 each, there will be one or two that are definitely superior compared to the others. The same factors hold for collecting in the field. Out of several dozen specimens collected there will always be a few that are obviously superior.

The point is to add the best available specimen from material collected or offered for sale, keeping it *within your budget* and size specialization.

Since the early seventies the subject of minerals as an investment has appeared in books, magazine articles, and dealer advertisements. This is an interesting and very realistic consideration. Fine specimens can be outstanding investments. There are countless examples of astonishing appreciation in specimen values as certain types of top-quality specimens become virtually unobtainable. What was once abundant on the mineral market very often disappears in a few years—and values escalate. For example, in the early 1950s between $15 and $25 would purchase an excellent small cabinet-size specimen of wulfenite from the Ahumada mine, Sierra de los Lamentos, Chihuahua, Mexico. The material was abundant, and an amazing proportion of the specimens were of first-class quality. A similar specimen on the 1970 market could easily bring $200. There are other examples, such as dioptase from a locality in the former French Congo, which sold for about $30 in the early fifties and now brings perhaps $400 or more. Fine kunzite crystals from the mines around Pala, San Diego County, California that sold for $100 in the forties might now bring $1,200. So it goes. Obviously, the rate of appreciation is better than most stocks!

Due to the increasing number of collectors and the limited number of superlative specimens, the law of supply and demand has definitely created an atmosphere of investment possibilities. However, you really better know what you are doing or be ready to accept unqualified, honest, and knowledgeable advice.

There is no question that many types of specimens have markedly increased in value as the years have passed, assuming that the specimen involved is of top quality. A fine-quality mineral specimen can usually be considered a good investment, provided that you are sufficiently familiar with the field and the current status of market availability to know if the asking price is a fair one. Unfortunately, there is no quick and easy course in ascertaining this: it is the product of experience, seeing shows and exhibits, keeping abreast of dealers' prices, and a certain degree of luck. Good buys are always available—it is the ability to distinguish them that causes the problem, like any other investment.

Determining the specific value of an individual mineral specimen is not as easy as it might seem. There are no official catalogs, price lists, or rule books covering mineral specimens. In this case the field differs tremendously from other avocations, such as collecting old firearms, antique dolls, glass paperweights, coins, seashells, rare books, or stamps. For these hobbies and many others there are regularly issued catalogs or listings that give well-established market prices.

What is really different about mineral specimens? Plenty. Every mineral specimen, good or bad, is individual and unique. No other specimen is exactly like it. A crystal-lined pocket in a mine may yield five hundred specimens of wulfenite, yet none will be precisely alike. Some will be larger or smaller than others, some attractive and others unattractive, some damaged and chipped and others free of blemishes. All are individuals—some in the category of top-quality specimens, most of them average, and some totally worthless.

Determining values involves becoming familiar with the nuances and variables that can affect mineral specimens. For example, a damaged crystal of very scarce native arsenic has little if any value. That the specimen is physically damaged is the key determining factor. On the other hand, a quartz crystal may be fantastically beautiful if it contains inclusions of pink kaolinite, green chlorite, and radiating bursts of golden rutile crystals! Although this example involves a very common mineral, a specimen of exceptional beauty has a far higher value than an ordinary quartz crystal. If you are trying to establish a dollar value on a specimen, keep in mind that *asking* a certain price and *getting* it are two different subjects!

Specialization

These considerations are particularly meaningful for the experienced collector who limits his collection to a certain specimen *size* or *specialization.* Even a multimillionaire would find it impossible to assemble a "complete" mineral collection. There are individuals who have invested hundreds of thousands of dollars in private collections; nevertheless, you do *not* need a great deal of money to build a truly superb collection. Finding a level of collecting suitable to your pocketbook is wise.

In essence, specialization in some aspect of the hobby can make a true expert out of an amateur collector. Many such collections not only have very stable monetary value but high scientific value as well.

Obviously, mineral specimens come in all sizes from micromount to large museum pieces. Perfection is always greatest or most common in small specimens. Consequently, the factor of size has a very real meaning for value—from the practical standpoint. For example, small, perfect, gemmy aquamarine crystals may be readily obtained from Brazilian localities in top-quality specimens. However, a *large* aquamarine of similar quality would be very costly, since it falls within the gemstone-mineral category and since large undamaged specimens are uncommon and difficult for the collector to obtain.

Similarly, there are many minerals that occur in superb small specimens but either do not occur in large specimens or else are unattractive in such sizes. For example, the common mineral dolomite occurs frequently encrusting rock matrix as very showy, *small* saddle-shaped crystals in various shades of pink, tan, or white. Thumbnail or small cabinet-size specimens can be very attractive, but a huge plate of dolomite is somewhat uninteresting. The combination of small-size crystals and a large mass of matrix may not yield as desirable a specimen as a much smaller matrix. Of course, large platelike specimens can often be broken or trimmed into desirable smaller sizes.

Keeping the basic quality of specimens in mind, coupled with relative size consistency and a definite organization or specialization, can result in a collection that is worth far more than a heterogeneous and motley assemblage of fossils, rocks, ores, and minerals.

Quality

There is little doubt that the most important value factor, regardless of the rarity or abundance of a particular mineral, is the overall *quality* of the individual specimen. Quality refers to the perfection and attractiveness of the specimen, or its basic physical condition coupled with a pleasing format.

Regardless of the kind of mineral, a clean, undamaged, showy crystallized specimen is the most desirable and commands the highest price. Specimens that are broken, damaged, cleaved, scratched, bruised, chipped, feathered, abraded, faded, or just dingy are not considered first-class material. Similarly, specimens that are etched, stained, coated with unattractive encrustations, or otherwise discolored are of little value. Such specimens may be interesting for study purposes, however, and certainly a great deal can be learned about minerals from very ordinary specimens.

If the mineral involved is crystallized, the overall quality of the crystals is very important. Fine single crystals, crystal groups, or crystal associations (two or more minerals crystallized on the same specimen) are the most popular type of mineral specimens, and the finest should show well-developed, sharp, easily defined crystal faces. Although many people specialize in loose single crystals, the most desirable specimens are those in which the crystals appear singly or in groups in an attractive format *on* a rock matrix.

Coupled with the all-important basic physical condition of the specimen is the aesthetic factor. Mineral specimens can be breathtakingly beautiful; they can be stunning to the point of being almost unbelievable. A great deal of the impact of such specimens rests with an outstanding format, which in turn generates aesthetic pleasure.

Beauty in minerals is difficult to describe, and what is beautiful to one person may not be to another. Nevertheless, a specimen with a pleasing format is highly desirable regardless of the size involved. Specimens in which the crystals are sprinkled or clustered in an attractive manner or dramatically located on the matrix are highly prized. Many times a collector encounters specimens that are well crystallized, bright, and showy yet are on a cumbersome, ungainly, or odd-shaped matrix and are simply not aesthetically pleasing. A good rule of thumb is, "Does the specimen display well?" In other words, can it be displayed with ease; does it show its physical form to good advantage? An awkward specimen containing very fine crystals is often almost impossible to display!

Some minerals occur in truly spectacular, beautiful crystal forms and colors, while others may almost never be found in attractive specimens. For example, azurite, a copper carbonate commonly found at many localities in brilliant deep blue crystals, is considered a beautiful mineral by most collectors. However, rhodozite, a very rare caesium mineral, occurs only in small, roundish crystals of cream or yellowish hue. There is really nothing attractive about them, but they are quite valuable. Consequently, in terms of value the beauty or appeal of a specimen is definitely a plus factor, yet always relative to availability. Needless to say, a specimen of azurite with several broken or damaged crystals has very little value—basic physical condition is always the most important consideration.

There are specimens on the market that have been "improved" by using acids to remove undesirable stains. It takes considerable experience to detect such man-induced changes on a poor specimen. At the same time, there are *some* superb types of specimens that are exclusively the result of controlled acid use. For example, California's beautiful blue benitoite crystals associated with jet-black neptunite on a soft white natrolite matrix are entirely the product of dissolving the massive natrolite around the benitoite and neptunite crystals with acid.

Why is overall quality rated so high? It is simple. Truly top-quality, virtually perfect, exceptional specimens are not abundant. A limited number of outstanding specimens will be created in an ore deposit in the first place! Mining activity—drilling, blasting, and loading—will damage or destroy many other potentially fine specimens. Comparatively small numbers of specimens that ever reach the so-called market are undamaged by handling, packing, shipping, and displaying. These three factors are always potentially working against the abundance of superb specimens. Also, mineralogy is unusual in that every specimen is, indeed, unique and one-of-a-kind.

Locality and Classic Specimens

A very realistic value consideration goes hand in hand with the abundance, or even current existence, of a given mineral at a particular *locality.* This is especially important to advanced collectors and especially to those who specialize in minerals from a particular locality, mine, or mining district. There are many collectors involved in this type of specialization, which in turn means that a very mediocre but acceptable specimen from a particular locality may be worth much more to one collector than to another.

The term *classic specimen* is often encountered in mineralogical literature. This refers to specimens from famous *old localities* or from a locality involved in the initial description of a particular mineral species. The early editions of the *Dana Textbook of Mineralogy* list worldwide localities of major importance for mineral species as they were first described and/or collected. Many of these classic old-locality specimens are extremely difficult to obtain and only become available when a private collection is sold or a museum becomes involved in a trade or disbursement of duplicate specimens.

There are hundreds of "the classics," all easily identifiable by the knowledgeable collector, since individual locality finds are incredibly distinctive. Specific localities produce specimens as distinctive of *that* locality as the fingerprint of a collector. It does not take too much time to become reasonably familiar with typical old-locality classics, such as the beautiful butterfly twin calcites or deep green fluorites from England, manganite from Germany, stibnite from Japan, epidote from Alaska, malachite pseudomorphs from South West Africa, rhodochrosite from Colorado, topaz from Siberia, native silver from Norway, stilbite from Iceland, sulfur from Sicily, tourmalines from California, wulfenite from Mexico, or cassiterite from Bolivia. Incidentally, old labels showing the past ownership or status of a specimen are definitely valuable assets to any specimen and should *always* be obtained when a specimen is acquired by purchase or exchange and *retained* with such specimens.

Fine specimens are usually encountered in mines and, except for gemstones, are completely incidental to the commercial purposes of the mine. For example, the gorgeous translucent cerise-red wulfenite crystals from the famed Red Cloud mine, Yuma County, Arizona are encountered as part of the ore body. No company would mine the deposit primarily for wulfenite specimens but for lead ore, and in this case the wulfenite is a legitimate part of the ore due to its lead content. When the mine tunnels, stopes, and shafts have passed through the areas of ore containing fine specimens, the source is exhausted.

As the history of mineral collecting indicates, there are dozens of old specimen localities that are no longer operating as working mines. Often the mine has moved through the crystal-producing areas, and the specimen supply is considered exhausted. Good specimens from such well-known localities can command a premium price by virtue of their unavailability! Fine crystallized specimens from the Mammoth-St. Anthony mine, Tiger, Pinal County, Arizona have brought astonishing prices in recent years, mainly because the mine has been closed since 1952 and chances are remote that the mine will ever be opened again.

Another interesting point is worth noting in these Mammoth-St. Anthony mine specimens. Frequently, a very common mineral will be comparatively scarce at a famed mineral-collecting locality. A locality may produce a dozen different superlatively crystallized minerals, but a couple of really common species may seldom be found at a particular site. At Mammoth, for example, fluorite is definitely uncommon, although in dozens of ore bodies it is a very common, well-crystallized mineral. Fluorite specimens from Mammoth, therefore, are of particular value to Mammoth collectors and bring prices far in excess of comparable fluorite specimens from other localities.

This mine is also an outstanding example of how complex an ore deposit can be and, in this case, how rich in both oxide and sulfide minerals. The mine has produced quality specimens of over fifty different minerals, including a number of very rare species. Many of these, such as wulfenite, vanadinite, and cerussite, are lead minerals; dioptase, azurite, antlerite, and malachite are coppers; and hemimorphite, willemite, and smithsonite are zincs. However, when the mine was first discovered and opened in 1879, it was operated for gold! The gold was recovered entirely from a wulfenite-vanadinite-type ore. In later years, as values increased for copper, lead, and zinc, the mine was operated for these materials as well.

Another example concerns minerals from a still active mining district that have not been encountered in similar specimens for many years. In the early 1900s the mines around Tsumeb, South West Africa produced a number of huge, flat azurite crystals of magnificent color and perfection. These beautiful crystals are the size of an outstretched hand. Crystals of this type, size, and quality have not been found since, although the various mines of the area continue to produce other azurite specimens as well as many other crystallized minerals. The value of the old Tsumeb azurites is high, and, like many unique finds at specific localities, they are absolutely unlike any other azurites, immediately distinguishable to the discriminating collector. These examples show that there are numerous cases where the specific locality is a very important, highly desirable, and expensive commodity!

Rarity and Abundance

Rarity is an important value factor. If a mineral is deemed truly rare and occurs in a specimen of exceptional format, perfection, and beauty, then its value is going to be very high. There are many individual specimens in various museum and private collections throughout the world that would be in the $50,000 to $100,000 bracket. There are also several minerals with comparatively few known specimens, and their value will be high if the specimen is of generally acceptable mineralogical quality. At the same time, there are truly rare minerals that are very unattractive and may be valuable only for the scientific-research collector or the advanced collector of mineral species.

In terms of abundance or availability, mineral specimens fall in one of these general categories: *rare*, *scarce*, *uncommon*, *common*, and *abundant.* There are hundreds of minerals within each category, but typical examples, in respective order, are: paradamite, manganite, smithsonite, calcite, and quartz. The basic abundance or scarcity of a given type of specimen definitely influences it value. Of course, the physical condition and quality of the specimen must also be good.

There are several interesting aspects of the rare-to-abundant scale. Some minerals that are common to uncommon in general occurrence are rare as *crystals!* A good example is turquoise, which is almost exclusively found in the massive form used for jewelry purposes. Very fine, tiny crystals of turquoise have been found and are considered a rarity. Similarly, there are some materials that a novice might assume are very common—but are not. For example, native lead crystals are very rare indeed. Despite the fact that lead ores such as galena are among the world's most abundant minerals, pure native lead in *any* form is very, very scarce: specimens, although unattractive, are quite valuable.

Unusual Crystals

There are many instances in which a particular locality will produce crystals that are either exceptionally large by comparison with previously known specimens or startlingly different from normal expectations. A good example is cuprite, a cuprous-oxide mineral that occurs in attractive ruby-red, isometric crystals. When it is found in hairlike, elongated microcrystals, it is known as chalcotrichite. Cuprite was usually found in comparatively small, modified crystals showing cubic, dodecahedral, or octahedral forms. Crystals ranging from 1/16" to large specimens showing ½" on a crystal face were expected in the finest specimens. This all changed in 1973 with the find of huge cuprite crystals encrusted with malachite at Onganja, South West Africa. Translucent ruby-red crystals were found as large as 3" on a face and of excellent development. These were true giants for this mineral, and their subsequent values were very high. The material can be faceted into stunning demonstration stones of vibrant purplish ruby-red, although it is too soft for normal jewelry use.

There are many aspects of the *unusual* that collectors become familiar with only by experience. It might be hard to become excited about a perfect ½" cubic crystal of boleite—unless you were a thumbnail collector and knew that this was a truly enormous crystal of a very uncommon mineral!

The real monsters of crystallography are seldom worth much as specimens, regardless of their great size. Most extremely large crystals are crude, imperfect, and unattractive. There are some exceptions, however, particularly in the case of quartz, which has been found in 4′ and 5′ specimens well worth museum display space.

Crystals, usually found "frozen" in a host-rock matrix complex, can reach gargantuan proportions. For example, beryllium mines have actually been opened in huge single beryl crystals in South Dakota! The Iceberg Claim, near Picuris, New Mexico, produced two minable calcite crystals weighing an estimated 25 and 35 tons, respectively! Galena, calcite, selenite, and fluorite have been found in crystals ranging from 1′ to 4′ on a crystal face. In general, such specimens are of very poor quality and are essentially curiosity items; however, a few exceptions have had ore value or have made their way into museum displays, and some are used as decor items in businesses and homes.

Popularity and Fads

Popularity is an interesting yet somewhat hazardous aspect of values. New mineral finds are made each year as new mines are opened, old mines reactivated, or operative mines work into new ore-body areas, producing specimen materials. Often outstanding finds are made of a new mineral at an old locality, or a well-established species is found in striking format at a new mine. New finds are always exciting and of constant interest to collectors. Often the new material is indeed limited—perhaps a few hundred choice items before the mine source works its way through the specimen-producing area. The specimens so removed may be all that will be found, or large quantities may be found in a few months—you never know! All you really know is that due to the basic dynamics of mining, no fine-quality specimen is limitless—the source will surely be depleted sometime. Infrequently, however, a huge deposit of material is encountered, which may actually deflate the value of a certain type of specimen. This is particularly true if the new material is of far superior quality or radically different from material already in collections.

For example, in early 1971 very choice botryoidal specimens of white, tan, peach, rose, and lavender smithsonite from a Mexican locality appeared on the market. The material was choice, different, and colorful and made a fine addition to any collection. It was purchased in depth. A year later the market was literally flooded with identical material at about one-third the cost of the initial specimens! By 1975, however, quality material was apparently no longer available—the market may have been saturated—so the value of the smithsonite from this particular locality was stabilized. This helps point out that the field of mineral collecting is very flexible; it bends with the wind in terms of the abundance and/or availability of a certain material. The very finest of the smithsonite under discussion will always be a choice collector's item and will also constitute an immediately identifiable locality, since the specimens are very distinctive.

Fads are another realistic feature of the value story. The history of gemstones in particular is subject to fads and can even represent a multimillion-dollar variable! For example, over the centuries opal has been in and out of fashion several times. At present, opal is enjoying a revival of popularity, and opal mining in Australia has increased markedly in consequence. In a similar vein, the appeal of garnets has fluctuated many times, considered either old-fashioned or the thing to wear as social, political, and fashion fads dictate. Fads and fashion go together and can have a profound effect on the gemstone and the mineral-decor fields. At present, there is an astonishing fashion-oriented upswing in the popularity of turquoise and related Indian jewelry. The sudden demand has exceeded supply and has produced a variety of techniques to "improve" low-grade turquoise or to fake it outright.

Mineral specimens per se are not as subject to these whims, yet particularly dramatic and spectacularly colorful new finds lend themselves to a wave of popularity. For example, at present wulfenite seems to be a mineral in particular favor with advanced collectors. It is immensely attractive and appealing in well-crystallized specimens from many localities. Much the same can be said of its related mineral, vanadinite. Similarly, the spectacular specimens of dioptase are very popular at present. Of course, this popularity was triggered by the various mines presently producing many superior specimens of wulfenite, vanadanite, and dioptase: the current availability factor is high. Perhaps they are popular because discriminating collectors know only too well that the supply of top-quality material can never last.

Stability

As mentioned earlier, it usually comes as a shock to the beginning collector to find that many mineral specimens are not "solid as a rock," and, as a matter of fact, many rocks are far from being solid! There are numerous types of specimens that are reasonably attractive when first mined but quickly lose their luster, turn opaque and powdery, and then actually disintegrate. They are unstable and, unless given specialized preparation attention, will disappear. Marcasite from certain localities is renowned for its instability, and realgar, orpiment, vivianite, laumontite, and pyrite from certain areas may also be very unstable. Obviously, *stability* is important: a specimen that is on its way out has little value.

Repaired Specimens

There is no general agreement between curators and collectors on *repairing*, yet it is a factor that all collectors should be well aware of.

Specimens do break during mine blasting, actual collection, preparation, or mishandling of a collection as the years go by. Various adhesives, of which epoxy is by far the best, can mend a specimen so that only very close inspection can reveal that it has been broken and recemented.

Repairing specimens is a standard procedure for many museum curators and seemingly does not constitute a very negative value factor. There are many well-known specimens in museums and private collections that have been repaired. Some of these have appeared as illustrations in a number of publications as well. As a matter of fact, a number of the finest and most dramatic large tourmaline specimens from both Maine and San Diego County, California, exhibited in collections the world over, have been repaired. In this case the damage was done in the original pocket in the host pegmatite rock. Tourmalines are found at these localities that have been broken naturally and then resealed themselves!

If a specimen has to be repaired, the mending job should be virtually flawless, or else the value is definitely decreased. Most private collectors prefer not to have repaired specimens in their collections, but sometimes it is unavoidable. If a repaired specimen is in a collection, the catalog listing should clearly indicate that such is the case. Similarly, dealers offering such specimens for sale should always include this fact on their label.

Cataloging and Labeling

There is no question that an accurate, thorough job of *cataloging* a mineral collection increases its value both scientifically and monetarily. Virtually any specimen is worth more if it is accompanied by a *label* giving the name(s) of the specimen and related locality data. Good cataloging and labeling can increase the value of a somewhat mediocre specimen beyond that of its basic physical appearance. Sometimes minerals from an unusual old locality may not be particularly attractive, but accompanied by a good label or card-catalog entry, the value is enhanced.

As the years go by, specimens often pass from collector to collector, gathering labels as they go. Some dealers make a point of advertising in their catalogs or listings that a given specimen includes a special label. There are numerous institutions and famous collectors whose collection label is definitely a prized aspect of the specimen. In a well-managed, advanced, private mineral collection it is not unusual to find a specimen accompanied by four or five labels tracing the specimen's past-to-present ownership status.

Self-collected Collections

There is one type of collection that is particularly difficult to evaluate—a collection of *self-collected* specimens. Many collectors specialize along these lines, and many of these collections are magnificent. You do not have to purchase specimens to have a top-quality collection. Many of the best gem- and mineral-society shows will often have an exhibit or competition category of self-collected minerals. You will usually be surprised at the quality of some of the displays.

The other side of this category is the modest collection that may have little if any dollar value but is priceless in terms of memories—a delightful treasure house in which to recall and reminisce on past experiences. Perhaps these are the finest of all specimens.

Gemstones

Precious or semiprecious *gemstones* automatically fall into a special value category that is reasonably obvious. Gemstone materials, especially if they are well crystallized, such as emerald, beryl, aquamarine, morganite, kunzite, topaz, and tourmaline, have values that are firmly established by the jewelry industry. In order to keep a superior crystallized specimen of gemstone quality away from the jeweler's faceting equipment, you must expect to pay a gemstone premium.

The term "gemstone" can also be very misleading. There are innumerable good specimens of gemstone minerals that have no commercial value as cutting material yet may constitute fine collector's specimens. In other words, crystals of aquamarine or topaz may be found that show fine crystal structure and are showy and bright yet are so flawed or cloudy that they contain virtually no faceting-grade material. Such specimens normally become good collection material. Nevertheless, crystallized gemstone minerals of good specimen quality always bring higher prices than many other categories of material.

There are good reasons for these higher values in addition to the cutting quality of the prospective material. First of all, good gemstone-type minerals are seldom common, and many would be considered rare. Secondly, most gemstone minerals are mined specifically for *that stone*, and, with the exception of certain gravel deposits, most of the time this is hard-rock mining.

Hard-rock mining for gemstones, such as in the pegmatites of Oxford County, Maine or San Diego County, California, is very hard work, expensive, and totally without any guarantee of success! Gemstones are elusive, and a mine may operate for months without striking a single gem-bearing pocket. Then, surprisingly, two or three may be found in a space of a few days. Gemstone mining is always a gamble, and this factor alone helps to keep the price structure high on quality specimens. For every ounce of quality tourmaline taken from the pegmatities of San Diego County, tons of totally worthless rock have been removed as well.

Gemstones are mined exclusively and primarily for the jewelry trade, and in essence this is the only reason for gemstone mines to be operative. The discovery of an overwhelming number of *other* attractive and desirable mineral specimens is entirely *incidental* to the mining activities. For example, the fine complex of wonderfully crystallized specimens from the mines around Tsumeb, South West Africa was found during basic mining operations for copper and lead ore. The specimens are incidental and are simply chemical components of the ore body. The mine is not run for these fine specimens, while gemstone mining is exactly the opposite.

There are always unusual exceptions in the gemstone category in which a comparatively common or even unattractive mineral is found at some isolated locality in gem-quality material. The mineral zoisite is a good example. Zoisite is usually green and is an undistinguished member of the epidote group of minerals. A pink variety called thulite has some decorative use. However, in 1967 a discovery was made in Tanzania of zoisite in well-developed transparent crystals of an astonishing violet-blue color. The material was named tanzanite. The best of these specimens can be faceted into very handsome gemstones and have enjoyed considerable demand since the discovery. There are numerous other examples of unusual isolated finds of normally nongemstone material occurring in gemstone quality. In many cases the mineral involved may be far too soft for practical use as a gemstone but may lend itself to faceting for museum or collecting purposes.

Precious Metals

Like gemstones, there is a basic intrinsic value attached to the *precious metals*, gold, silver, and platinum. Even this prime factor can be incredibly increased if the specimen is well crystallized or occurs in sheets or wirelike forms. This is often the case with gold or silver. Superb specimens that may have perhaps $250 worth of actual gold by weight may occur in such a striking format that they might bring $3,000 apiece.

As a rule, specimens that clearly show the precious metals command substantial prices. Similarly, there are some minerals in the silver family, such as proustite or argentite, that are considered quite valuable when found well crystallized.

Incidentally, the timeworn adage "worth its weight in gold" seemingly implies an ultimate value. Actually there are many minerals found in fine crystallized specimens that are worth far more than their weight in gold!

Man-made Crystals

Various processes invented by man produce *artificial*, or *synthetic*, mineral materials for various purposes, such as industrial-grade diamonds, quartz crystals for communications technology, and synthetic emerald, ruby, and sapphire gemstones, among others. All of these have perfectly legitimate uses, provided (especially in the case of synthetic gemstones) they are marketed as just that—synthetics! However, a number of colorful chemicals can easily by crystallized by man on specimens of ordinary rock serving as the host matrix. Usually these are so unreal or gaudy in appearance that they may be detected easily and rarely fool a reasonably knowledgeable mineral collector. Another problem is that many of the most spectacular crystallizations are quite toxic.

Artificial man-made chemical crystals do not belong in a fine mineral collection. They are curiosities. They have no authentic mineralogical value but may be of interest to the crystallographer.

Man-altered Crystals

This category involves an experienced gemologist or mineralogist, since many natural crystals of various minerals can be *altered* drastically in color by various atomic-radiation processes. It is exceptionally difficult to ascertain the difference between a genuine natural yellow diamond and one that has been turned yellow by radiation. Similarly, crystals of transparent quartz can be turned into very showy specimens of smoky quartz by exposure to radiation. A fine plate of smoky-quartz crystals brings a higher price than a clear and transparent plate.

Fakes

Yes, like virtually anything in which man expresses interest, there are deceptive practices and *fakes* in the mineral field—believe it or not. Some are quite ingenious.

Since fine crystals on a rock matrix have always been the most sought-after type of mineral specimen, the art of gluing loose crystals to a rock matrix has been practiced for years! Tourmaline "on matrix" was a particular favorite. Some of the false pasteup jobs are distinguishable with normal scrutiny, since they often look almost "too good to be true" (crystals spaced out just a little too perfectly!).

Other feigned specimens can match the finest examples of the counterfeiter's art. One of mineralogy's classic specimen types is the combination of attractive pink octahedral crystals of transparent fluorite perched atop terminations, or clustered on the prism faces, of water-clear quartz crystals from selected localities in the Central Aar Massif granites of Switzerland. They have always commanded a high price. During the early 1950s some superb fakes of octahedral pink fluorite glued to Swiss quartz crystals reached the market. The original pasteup source was believed to be Germany. The fakes were so perfect that at least two major dealers and a couple of museum curators were fooled. As the months went by, the formerly clear adhesive began to darken slightly and exposed the specimens as fraudulent.

More recently, during the early 1970s there have been many cases of faked native-gold nuggets and crystallized gold. These shams have been prepared in sealed boxes with a transparent lid revealing the "gold" specimen mounted for viewing. The purported gold is nothing more than a rock fragment or a plastic casting on which gold paint or gold leaf has been applied. The bottom of the box was weighted with lead to give the specimen the appropriate specific gravity! Again, many unsuspecting buyers have been fooled by this ruse.

Early in 1975 some remarkable faked emerald specimens from Columbia appeared on the market. At the famous emerald mines the green crystals are found in veins of white granular calcite associated with a fragmented dark gray rock matrix. Apparently, both the rock matrix and the calcite had been ground up, mixed with epoxy, and cleverly used to sculpture a very realistic matrix for small, loose crystal fragments of emerald, which were embedded in the material. The result is a specimen that almost requires a microscope to detect the fraud.

Fortunately, the opportunity for passing off successful fakes in the mineral field is relatively small; nevertheless, it is wise to remember the phrase "caveat emptor"—let the buyer beware.

Fossils

In recent years there has been a marked increase in the number of amateur paleontologists, and good-quality specimen material is readily found in almost all states.

Fossils constitute our only record of the ascendance, domination, and extinction of prehistoric life forms; consequently, many of the exceptionally rare specimens are scientifically priceless. These are museum-display items or specimens for paleontological-research laboratories. Fortunately, modern technology can create exact replicas for study purposes. Incidentally, many very valuable fossils may not look like much or seem particularly interesting.

For the amateur collector and hobbyist there are numerous well-established clubs and societies devoted to this science, and values are often discussed on field trips and at meetings. There are also several reputable companies devoted exclusively to selling fossils to educational institutions, museums, and hobbyists. At least four of these firms produce very detailed, scientifically accurate catalogs and periodic listings of hundreds of fossils available from their stock with appropriate prices. As you would imagine, price is essentially keyed to the abundance and/or rarity of the particular fossil, followed by the degree of physical perfection.

There are countless fossil specimens for which the price structure is very low—$1 or less will purchase a good example. Beyond this are many specimens particularly sought by collectors due to their degree of perfection and appeal as a fossil form. Examples might be perfect fish imprints or skeletons, turtles, or oredon skulls. Such items can bring $50 to $500 for choice specimens. Superb agatized pine cones from Patagonia in Argentina may bring $100 to $400.

Like choice mineral specimens, the supply of an available fossil may be very limited—which in turn directly affects its value and price. A deposit in Oregon of well-fossilized nuts has yielded odd and choice items that have brought substantial prices.

The ardent mineral collector should remember that fine fossils have a good trade value, either with a dealer or with another collector. A good way to expand a mineral collection is to trade fine fossil specimens; a fossil collector may collect and exchange good minerals for fine fossils.

In any event, consulting dealers who specialize in fossil materials and reading their current catalogs are the best guidelines for values.

Pseudomorphs

Many collectors specialize in *pseudomorphs* or have extensive suites of such specimens within a general collection.

The value of pseudomorphs is quite difficult to place on a scale. Outstanding examples, such as gypsum pseudomorphs after glauberite from Verde Valley, Arizona, may be purchased for perhaps a dollar—or several hundred dollars may be spent for a fine malachite pseudomorph after azurite from Tsumeb, South West Africa. That is quite a span!

Perhaps half the fun of collecting pseudomorphs is attaining the sophistication in mineralogy to recognize a pseudomorph when you see one! In any event, there are many famous localities for pseudomorphs that are well known to advanced amateur collectors: quartz after barite from Colorado, chalcedony after aragonite from Mexico, native-copper "skulls" from Michigan, malachite after cuprite from France, native copper after aragonite from Bolivia, and quartz casts after anhydrite from New Jersey are some examples.

Since many pseudomorphs are not showy, they are usually of more value to the specialist than to the average collector. Unless the specimen is of outstanding beauty or a gemstone, pseudomorphs do not have unusual dollar values.

Variables

There will always remain the final and ubiquitous category that can only be referred to as *variables*. No matter how collections or relative values are categorized, there will always be someone who will say, "Yes, but I collect such-and-such." Such factors are one of the most stimulating aspects of this fine cluster of earth-science hobbies. For example, I have emphasized the merits of a basic size of specimen upon which to concentrate, and perhaps a particular specialization as well. There are those who will disagree, and this is understood. I can think of a fine and very valuable private collection on the West Coast that is composed of several hundred top-quality specimens displayed in custom-designed, built-in, illuminated glass cases. The specimens range in size from 1″ × 1″ to 10″ × 12″ and in weight from an ounce to over 20 pounds! In other words, this collection embraces virtually all sizes of specimens and is completely nonspecialized. The value and charm of the collection rest on two factors: first, for the most part the specimens well fit the quality categories discussed in this chapter, regardless of size; and second, they are displayed with exceptional *artistic* talent. This particular collector is essentially interested in quality mineral specimens that are aesthetically pleasing. One or two are truly rare, several are uncommon, and most are fine examples of good-quality, well-crystallized mineral specimens. This collection is certainly enjoyed by all who see it. To each his own.

Many other illustrations of elusive variables can be mentioned. Consider the valued acquisitions that a knowledgeable collector will purchase or exchange for regardless of fundamental quality. For example, most collectors have specialties within their collections, and there are certain specimens that they have been seeking for *years*. Usually these are old-timer specimens from long exhausted or very limited localities. When these become available via the disposal of an old collection or museum duplication, they are rarely first-class, choice specimens. However, they may be the best that the collector will ever have the opportunity to obtain, and, realizing this, he will purchase a less than choice specimen simply because it is a long-standing item on his want list.

Humorous as it may seem, some specimens are purchased simply because the buyer is fascinated by some particular aspect of it. He simply *likes* it. This is called emotional buying in other collecting fields; however, I feel that everyone should collect what *he* enjoys.

Many collectors, both amateur and advanced, often feature attractive displays of minerals in their homes, and, consequently, some purchases may be based simply upon a specific color, size, and format to fit a particular spot in a display case. Sometimes a dealer will be approached with the question, "What have you got in a nice crystallized specimen, about 3" × 4", and in red?" This is an aesthetic-type inquiry. Incidentally, fine-quality crystallized minerals that are red are not common!

In recent years there has been a major market for fairly large, colorful, and aesthetically pleasing mineral, rock, and fossil specimens for use in the home or office purely as decor items. If homes display paintings, ceramics, and metal sculpture, why not use a large, showy group of selenite crystals as well?

Trading and Selling

Obviously, mineral dealers get their stock of material from somewhere, and dealers are seldom prime collectors of specimens per se. Consequently, dealers are always looking for contacts with miners or collectors to supply them with good specimens of minerals, fossils, or lapidary materials. There are dealers who specialize only in wholesaling specimens to retail dealers.

Many collectors may realize a modest sum by selling self-collected specimens to dealers, but don't expect to make a financial killing by doing this. Dealers have to make a substantial profit in order to stay in the earth-science-materials business—and the hobbies need good, reliable dealers.

If you are interested in expanding a private collection and have a limited budget, you might consider trading with selected dealers. Trading specimens for specimens—taking materials from the dealer's stock in exchange—will usually give the seller (trader) better dollar value. The dealer may be a bit more flexible on an exchange than with outright cash.

As an example, I have often traded with other collectors on the basis of the factors discussed in this chapter. Similarly, many deals have been completed with dealers, usually to my apparent disadvantage in terms of the number of specimens given up in order to acquire a desired specimen. However, there is another side to this. Let's assume that I have traded one hundred specimens of barite from the Palos Verdes locality in California for one specimen of dioptase from Tsumeb, South West Africa. Crazy? Perhaps not.

First of all, not being in the mineral business, there was no need to keep dozens of duplicate specimens for years. Also, they were collected in the field for trade or exchange purposes; the finest were retained for my own collection. I am interested in adding choice items that I do not have to my collection, and there will always be fine specimens for which I would never have the *cash* to purchase. In this case the dealer may have given me $1.50 per specimen in trade, which in turn allowed me to "purchase" a $150 specimen which otherwise I could not have afforded. At least this is a rationalization that keeps both the dealer and the collector happy.

It might be apropos here to comment on buying specimens through the mail from dealers. Always have an understanding *in writing* that any specimens purchased may be returned in a few days if they are not satisfactory for your needs. There are dozens of reputable dealers who simply would not conduct business otherwise. The point is that a given specimen may be described in perfectly honest terms in a catalog listing or advertisement and yet when examined may not meet your glowing expectations.

Let me give you an example from my own collection: specimen 2467, "Calcite, sphalerite, and dolomite, Bromley Hill, Nenthead, England (University of Bristol label 140-a), 2½" × 2¼" × 3", matrix encrusted with several translucent calcite crystals, black resinous sphalerite crystals up to 1", on china-white dolomite crystals."

Sounds pretty good, doesn't it? It is obviously an old-timer from a locality that most mineral collectors are anxious to have represented in their collections, and it sounds quite attractive. Forget it! All the things said about the specimen are quite true, but upon inspection you would find that it is almost impossible to display in an attractive manner! In other words, the crystals, which are good, are scattered all over and around the entire matrix, and there is no way to display the specimen. It is an interesting specimen and has value in terms of the old locality and the study merit of the well-developed crystals, but if you purchased this through the mail expecting a fine *display* piece from an old British locality, you'd probably send it back.

BIBLIOGRAPHY

The following publications have been used as reference sources in preparing this book and/or are considered by the author as particularly outstanding and useful items for further reading.

Books

Ash, Sidney R. and David D. May, *Petrified Forest: The Story Behind The Scenery*, Holbrook, Arizona, Petrified Forest Museum Association, 1969.

Bancroft, Peter, *The World's Finest Minerals and Crystals*, New York, The Viking Press, 1973.

Dana, Edward Salisbury, *A Textbook of Mineralogy with An Extended Treatise on Crystallography and Physical Mineralogy*, 4th ed., rev. and enlarged by William E. Ford, New York, John Wiley © (1898), 1922, 1949.

Dana, Edward, James Dwight, and Edward Salisbury Dana, *The System of Mineralogy*, 7th ed. 3 vols.; vols. I and II by Charles Palache, Harry Berman, and Clifford Frondel; vol. III by Clifford Frondel; New York, John Wiley © (1892), 1944.

Desautels, Paul E., *The Gem Kingdom*, New York, Random House, 1970.

———, *The Mineral Kingdom*, New York, Madison Square Press, Grosset and Dunlap, 1968.

———, *Rocks and Minerals*, New York, Grossett and Dunlap, 1974.

English, George Letchworth and David E. Jensen, *Getting Acquainted with Minerals*, New York, McGraw-Hill Book Co., 1958.

Fenton, Carroll and Mildred Adams Fenton, *The Fossil Book*, New York, Doubleday & Co., 1958.

———, *The Rock Book*, New York, Doubleday, Doran & Co., 1940.

Fleischer, Michael and John S. White, Jr., *Glossary of Mineral Species*, Bowie, Maryland, Mineralogical Record, 1975.

Halliday, William R., *American Caves and Caving*, New York, Harper & Row, 1974.

Hurlbut, Cornelius S., Jr., *Minerals and Man*, New York, Random House, 1970.

Jensen, David E., *Getting Acquainted with Minerals*, New York, McGraw-Hill Book Co., rev. ed., 1958.

Murdock, Joseph and Robert Wallace Webb, *Minerals of California*, Centennial Volume (1866–1966), California Division of Mines and Geology, Bulletin 189, 1966.

Northrop, Stuart A., *Minerals of New Mexico*, Albuquerque, New Mexico, University of New Mexico Press, rev. ed., 1959.

Pough, Frederick H., *A Field Guide to Rocks and Minerals*, Boston, Houghton Mifflin Co., 1953.

Ransom, Jay Ellis, *Gems and Minerals of America: A Guide to Rock Collecting*, New York, Harper & Row, 1975.

Roberts, Willard L., George R. Rapp, Jr., and Julius Weber, *Encyclopedia of Minerals*, New York, Van Nostrand Reinhold, 1974.

Sanborn, William B., *Handbook of Crystal and Mineral Collecting*, Mentone, California, Gembooks, 1966.

Shelton, John S., *Geology Illustrated*, San Francisco, W. H. Freeman and Co., 1966.

Sinkankas, John, *Gemstones of North America*, vols. I and II, New York, Van Nostrand Reinhold, 1959, 1976.

———, *Mineralogy for Amateurs*, New York, Van Nostrand Reinhold, 1964.

———, *Prospecting for Gemstones and Minerals*, New York, Van Nostrand Reinhold, 1970.

Sinotte, Stephen R., *The Fabulous Keokuk Geodes*, vol. I, Des Moines, Iowa, Wallace-Homestead Co., 1969.

———, *Geode Country*, Portage, Michigan, The Geodes Co., 1972.

Weibel, Max, *A Guide to the Minerals of Switzerland*, New York, John Wiley & Sons, 1966.

Periodicals

There are several magazines, journals, and periodicals in the fields of mineralogy, paleontology, and the lapidary arts. These vary from the very simple to those that are of a purely scientific nature. The following three magazines are all well recommended for both casual and advanced collectors.

Gems and Minerals (yearly subscription, $6.50; single issue, $0.70)
The official publication of the American Federation of Mineralogical Societies. This is a good general all-around magazine featuring articles for the hobbyist and layman on minerals, lapidary arts, and some aspects of paleontology. Contains many advertisements per issue.

The Lapidary Journal (yearly subscription, $6.96; single issue, $0.70)
An excellent magazine with emphasis upon the lapidary and jewelry arts, related technology, and field trips. The back issues comprise an astonishing compendium of procedures, tips, and how-to-do-it in the fields of cutting, polishing, and jewelry manufacturing. Contains hundreds of advertisements per issue.

The Mineralogical Record (yearly subscription, $10; single issue, $2; published bimonthly)
An absolute "must" for the serious beginning mineral collector as well as the advanced, sophisticated hobbyist. Features astute articles on minerals, famed localities, highlights of outstanding new finds in minerals, etc. Articles vary from highly technical to easily understood by an interested and serious collector of minerals. Back issues may well become collector's items by themselves. Contains advertisements essentially keyed to the collector of fine-quality crystallized-mineral specimens.

INDEX

Numbers in *italics* denote illustrations.